On the Metaphysics of Experimental Physics

On the Metaphysics of Experimental Physics

Karl Rogers

First published 2005 by
PALGRAVE MACMILLAN
Houndmills, Basingstoke, Hampshire RG21 6XS and
175 Fifth Avenue, New York, N. Y. 10010
Companies and representatives throughout the world

PALGRAVE MACMILLAN is the global academic imprint of the Palgrave Macmillan division of St. Martin's Press, LLC and of Palgrave Macmillan Ltd. Macmillan® is a registered trademark in the United States, United Kingdom and other countries. Palgrave is a registered trademark in the European Union and other countries.

ISBN 1–4039–4528–4 hardback

This book is printed on paper suitable for recycling and made from fully managed and sustained forest sources.

A catalogue record for this book is available from the British Library.

Library of Congress Cataloging-in-Publication Data
Rogers, Karl, 1967–
   On the metaphysics of experimental physics / Karl Rogers.
      p. cm.
   Includes bibliographical references and index.
   ISBN 1–4039–4528–4
      1. Physics–Experiments–Methodology. 2. Physical measurements.
3. Experimental research. I. Title.

QC33.R594 2005
530'.072'4–dc22   2004051401

10   9   8   7   6   5   4   3   2   1
14   13   12   11   10   09   08   07   06   05

Printed and bound in Great Britain by
Antony Rowe Ltd, Chippenham and Eastbourne

For Thomas

"*Hephaestus, the God of Fire, has become the supreme master of the world. His furnaces are roaring. He has dispelled the clouds of Asiatic mysticism which obscured his native mountain. He has girdled the world with hoops of steel. In plain unmetaphorical language this is the age of science, of machinery... Every weapon, every machine is the embodiment of human thought and purpose. The user adopts that thought and purpose, and behold – the machine has found its soul.*"

E.E. Fournier D'Albe, *Hephaestus or The Soul of the Machine* (London: Keegan-Paul, 1926), p. 1

Contents

1
Entering the Cave of the Shadow Puppeteers

The purpose of this book is to show how physics has been presented as a "natural science" and an "empirical science" on the basis of a hidden operational metaphysics. This metaphysics has allowed the traditional realist and positivistic philosophies of science to completely neglect the role of technology in the development of theories and observations within physics. Scientific instruments, such as telescopes and microscopes, are assumed to simply increase our perceptual possibilities and see what is "out there". The use of detectors, such as X-ray scanners, electron microscopes, and the Geiger counter has supposedly allowed us to understand phenomena in the visible world in terms of otherwise invisible entities. According to the traditional view, the practical value of such instruments clearly "proves" that science has made considerable advances, progressed, and technology has no further relevance to the philosophy of science. However, as I shall argue in this book, technological innovation has not only made new observations and experiments possible but it has also transformed our experience and conception of reality. Using a microscope or a Geiger counter does not merely involve seeing or detecting what is there. One must interpret the behaviour of the instrument in terms of an understanding of how it works. Making an observation using novel instruments is bound up with making novel techniques of representing what one sees and how the instrument works. These techniques are ordered into procedures and operations within a technological framework that orders how we use and understand the instrument. Furthermore, these instruments did not fall from the sky ready-made with an instruction book. They were innovated as a result of complex labour processes and protracted efforts against a historical background of expectations, challenges, demands, and the results of previous research. Technological innovation makes new research, observations, and representations possible; it also brings with it new challenges to achieve the anticipated possibilities of future innovation and investigation. It transforms the trajectories and content of research within physics. New machines produce new phenomena, data, and change how humans understand the world. The history of modern experimental physics is just as much a history of the innovation of new machines, instruments, and techniques, as it is a history of ideas, theories, and discoveries. New technologies create observational possibilities and conceptions of the criteria for the possibility of possessing knowledge of natural laws and mechanisms.

In this book, I shall describe and analyse how practices and experiences are emergent, related, and unified within a technological framework, by examining the ways that the interactions between beings in the world have been understood and communicated within experimental physics in technological terms. Experiences and practices must be made public in the form of explanatory representations if they are to be successfully communicated and reproduced. Public representations can take many forms, such as textual or verbal accounts, drawings or diagrams, apparatus or instruments, photographs, simulations, and models, and do not only take the form of mathematical theories. Profound changes in our ability to represent and explain experiences and practices involves the development of new conceptions of what it is possible to cognate, represent, manipulate, and control. The way that physicists describe and explain the performances of the experimental apparatus in terms of invisible structures and interactions, such as atoms, waves, particles, forces, fields, laws, coupling constants, or whatever, does not passively arise from the physicists' experience of the flashing lights, graphical outputs, and changing read-outs of detectors, instruments, and machinery. These descriptions and explanations require fundamental conceptions of what the responses of the apparatus mean in order for them to be unified into an experience of the effects of invisible entities, laws, and mechanisms. This is not only peculiar to subatomic or quantum physics. We all have experiences of objects falling to the ground when dropped and staying on the ground when left there, but there is nothing passively immediate in these experiences that allows us to unify these experiences as the consequences of mass, inertia, and gravitation, let alone being able to equate and unify these experiences with the movement of the moon and sun across the sky, or the shape and metric of a space-time continuum! We also would not be able to understand and have experiences of the effects of electricity and magnetism in terms of electromagnetic fields, charges, currents, and voltages, without fundamental conceptions by which we could equate and unify our experiences of the behaviour of otherwise distinct entities within the world. There is nothing innate in our ability to conceptualise lightning as an electrical phenomenon akin to the behaviour of an electrical spark between two charged plates. New experiences and discoveries in science require new conceptions. In order to understand how discovery is possible and made in experiments, we need to understand how these conceptions are made possible. Understanding how physicists use technological objects to innovate and understand the representational aspect of "empirical" description of novel phenomena, through interventions, visualizations, and models, reveals the complexity of the relationship between experiences and practices in experimental physics. Observation is an activity within a framework of interventions, interpretations, expectations, possibilities, and purposes. The artifice in designing, building, using, and interpreting novel technological instruments to make interventions is a particularly important feature of the novel dimension of scientific discovery. In order to explicate how experimental physics has been metaphysically conceived as a natural and empirical science, I shall describe how the technological objects within experiments can acquire an agency of their own and, as the means to innovate and disseminate new experiences, techniques, and instruments, become autonomous. I shall show how techniques and instruments are metaphysically understood and represented as a neutral means to experience and explain the facts of objective reality, and how the phenomenological experience of their autonomy is possible.

The cave of the shadow puppeteers

In the seventh book of Plato's *The Republic*, Socrates presented the famous cave analogy. In this analogy we are asked to imagine people living since childhood in a cave deep underground. Their necks and legs are tied to fix them in place and prevent them from turning around. A fire burning far above and behind them provides light and they can only see the cave wall directly in front of them. Between the fire and the prisoners is a walled path along which puppeteers carry puppets of people and other animals made out of stone, wood, and every material. Some of these puppeteers are talking, whilst others are silent. The tethered people are only able to see the shadows of the puppets projected by the light of the fire. The echoes of the voices of the unseen puppeteers bounce off the cave wall and seem to come from shadows themselves. If the prisoners could speak then they would think that the names they used applied to the shadows passing in front of them and that the shadows were truth and reality. Perhaps they would grant honours and praise to those who were sharpest at identifying the shadows as they passed by and were best at remembering the order of the procession. Perhaps they would consider those who devised some method for predicting the sequence of the shadows as being their wisest and most knowledgeable.

Socrates asked us to consider what would happen if one of these people were to be freed of his bonds and cured of his ignorance. Imagine that this prisoner was compelled to stand and turn his head, walk, and look up towards the light of the fire. Imagine him dazzled and pained by the light, unable to see the puppets whose shadows he had seen before, unable to make sense out of the voices that surround him. If he were told that his previous reality was comprised of only shadows and echoes and that he was now turned towards and closer to truth and reality, what do you think that his response would be? If he were shown each puppet in turn and ordered to tell us what it is, this poor man would no doubt be at a complete loss and would more than likely cling to his familiar, previous truth and reality. If he were forced to look at the bright and burning light of the fire then no doubt his immediate inclination would be to flee back into the comfort of the gloom and shadows. Imagine that he was prevented from doing this and dragged, kicking and screaming, up the rough steep path, past the puppeteers, past the burning fire, and out of the cave. Imagine that he was dragged up to the surface by force and hurled into the sunlight. Imagine him blinded by the light of the sun, unable to see a single thing that is now said to be true and real. He is stood, with his hands shielding his eyes, squinting at the painful and bright light that fills his vision. Of course, some considerable time would be needed before he could see the world outside the cave. At first he would only be able to see most clearly at night. He would be able to see the moon and the stars, as well as the shadows and silhouettes of things in the moonlight. During the next day he would be able to see shadows cast by the sunlight, then reflections cast into water, and only later would he be able to make out things directly. Finally he would be able to see the sun itself and study it. This man might readily infer, as Socrates suggested he might, that the sun governs everything in the visible world and is the cause of all the things that he has seen. Socrates then considered what the man would think if he remembered the cave, the prisoners, and the nature of truth and reality there. Would not the man think himself wiser now and pity the prisoners below? Would he not think himself in a happier condition than even the most honoured and praised of the prisoners? Would he not rather suffer anything than return to the cave, share their opinions, and live like that again? Socrates asked us to imagine what would happen if the man decided to return and help his former fellow prisoners. Imagine him returning into the darkness. His vision would now be dim in comparison with the prisoners. Would they not claim that his journey up to the surface had ruined his eyesight? Would they not ridicule him and the very idea that it was good to even try to travel to the surface? If he tried to free them then would they not resist or even try to kill him?

Socrates used the cave analogy to elucidate the relationship between the visible and the intelligible realms. The visible realm was to be likened to the cave. The journey to the surface was akin to the soul's journey to the intelligible realm and the study of the sun was likened to the difficulty in coming to see and know the form of the good and rational life. Such knowledge could only be achieved with considerable difficulty. For Socrates, it was quite unsurprising that those who have grasped the form of the good and rational life should have great difficulty and be open to ridicule when they are compelled to contend in the courts about justice and laws with those who have never seen justice itself. How could we expect any society to be just when its rulers do not understand or care what justice is? In discussing the education of the ideal city's rulers, the philosopher-kings or guardians, Socrates proposed that number, calculation, and geometry (used within every craft and science, including physical training, poetry, music, and astronomy) are the subjects that have the greatest power to awaken intelligence in children and are also useful for the art of war (enabling the rulers to be adept at commanding the city's defence). For Socrates, the study of the science of mathematics was a good for its own sake, as well as turning the soul's intellect to the study of the form of truth, and it sharpened the intellect to grasp the forms of justice, beauty, and goodness through the dialectics of reasoning and argument, which was necessary for the purpose of understanding the good and rational life. Education should not be considered to be merely a process of instruction, putting knowledge into souls that lack it, like putting sight into blind eyes, but instead should be considered to be the process of turning the whole soul to see what is already within it. It should be the craft of redirecting the sight of the soul to bring out the knowledge of the good and rational life that is already there. Socrates termed the educational process, which brought people from the shadows into the light, to be the ascent to true philosophy.

What has this to do with modern science? For the traditional philosophers of science, such as Bertrand Russell, Alfred North Whitehead, Rudolf Carnap, and Karl Popper, the theories of a modern science such as physics are straightforwardly representations of "the Laws of Nature" and experiments provide "empirical data" to verify or falsify those theories.[1] The objects of thought for these philosophers of science are the images, models, diagrams, theories, propositions, and hypothesis produced by working experimenters. These philosophers ignore the technological processes involved in the production of theses objects and presume that the results of experimentation, reported as "empirical regularities", "observations", or "data", are simply "the facts" against which the empirical adequacy of theories are to be tested. They presume that the abstractions, techniques, and material practices used in real experimental work are quite simply the neutral and transparent technological means to make observations and disclose "the facts". For these philosophers, science is primarily theoretical, whilst experimentation is merely a neutral technology to empirically determine which theory "fits the facts" better than the others. They all agree that physics successfully achieves progress in empirical accuracy and predictive power. They all agree that the representations provided by physics are "closer to the truth" or "more logically consistent" about Nature than previous representations, that science has lead to a growth in knowledge. Disagreements between these philosophers are concerned with the character of the scientific method by which representations are to be rationally ordered and justified by the empirical data. Some declare themselves to be positivists and tend to focus the debate upon which series of logical operations best characterise the general (and universal) principles of a scientific method. Others declare themselves to be realists and tend to reiterate the centrality of causal accounts and explanations to the intelligibility of the scientific enterprise. For the positivist, the scientific method is primarily a logical process of justification of the choice between theories on the basis of the facts, whereas, for the scientific realist, it is primarily a logical process of discovery and the role of theory is to explain the facts. Both kinds of philosopher agree upon what the facts are and leave it to scientists to provide them, whilst they focus their debate upon the nature of scientific theory and methodology. These philosophers maintain a clear and absolute distinction between theories and experiment, keep their backs turned towards the experimenters, and ignore how representations are actually produced and used in experimentation. They are akin to the prisoners in the cave analogy, sat with their backs to the shadow puppeteers, waiting for new shadows to be projected upon the cave wall, whilst the unseen puppeteers work with all kinds of materials to provide representations and facts for the philosophers' scrutiny. The philosophers argue and debate, using all the rhetorical and analytical skills their discipline provides, about which epistemology provides the best means to decide between theories on the basis of the representations and facts provided by the experimenters. They argue and disagree about whether any knowledge of a causal reality behind the representations is possible, whilst they agree upon the progress that theories have in matching predictions with the representations.

Since Michel Foucault and Thomas Kuhn published *The Order of Things* (1961) and *The Structure of Scientific Revolutions* (1962), respectively, the central role of dominant social consensus, authority, and power in the definition, content, ordering, and dissemination of scientific knowledge and the representation of reality has been discussed and analysed at great length.[2] Science has been studied as a culturally situated phenomenon in which the character of society transforms the nature of the scientific method, knowledge, and the representation of reality.[3] These critics of the traditional philosophy of science have tended to be extremely critical of attempts to demarcate science as a special rational activity. They have tended to argue that scientific work forms disparate, heterogeneous, fragmentary cultures of loosely associated sources of authority, technologies, research programmes, and personalities. They have attempted to place the fragmentary constellation of "sciences" into historical and social contexts in order to show that scientific knowledge, observations, practices, and values are completely contingent constructions and instruments of authority and power. It is the notion of the historical emergence of science from within the political and social contexts of Western civilization that has opened the very notion of "scientific rationality" to criticism.[4] In the light of this historicism, these writers tend to be critical of the traditional distinctions between nature/human, natural /social, fact/fiction, and truth/power. Science is often examined negatively in the light of its role in the social construction of modern human existence and knowledge. As part of the process of undermining the traditional view of science, critics attack the foundations upon which "science" has gained authority and power, and, as a consequence, they tend to be concerned with scientific organization, legitimization, authority, and justification, rather than theories, facts, discovery, and truth. This tends to run contrary to the epistemological individualism dominant in the traditional philosophy of science, because these critics emphasise the social and historical processes involved in the public development and acceptance of new theories. Rather than arriving at a new scientific theory in an intuitive flash of individual genius, as many of the traditional philosophers presume, the situation is more of a complex social struggle taking years, decades, or even centuries. Furthermore, scientists are involved in an industry of economic exchanges within the society in which science is emergent, producing values and powers for the wider world of commercial, military, and political ambitions. Whilst scientists may well maintain that they make and use their experiments and theories purely for the purpose of representing reality, in the wider world they are used as weapons, tools, products, and displays of national or civic prowess. Scientists receive materials and other resources for the production of machine prototypes. This consideration raises the centrality of social and cultural values to conceptions of "rationality" and "progress" in the acceptance and justification of scientific work. Traditional philosophers of science tend to consider values to be subjective or psychological and, therefore, can only be objects for rational judgements rather than constituents. They maintain a clear distinction between the "pure" science within scientific research and the "applied" science in the wider world. How the people in wider society use the products of scientific work is not the responsibility of scientists, claim these philosophers, because science is evidently value-free. The critics of the traditional philosophers have taken considerable pains to show example after example where this is evidently not the case, and critically examined the social function of scientific research within the political and economic structures of the wider society. They have also taken considerable pains to demonstrate how the very character of science is transformed by the demands of the wider world in which science is emergent and integrated. The critics of the traditional philosophers of science have examined that way that the rules by which the representations, the shadows on the cave wall, have been ordered, related, and analysed has changed from era to era. These critics have also sat in the entrance of the cave and watched how the philosophers' debate and interpretation of the shadows has reflected the prejudices of those philosophers as members of a wider society outside the cave. They have also watched how the choice of puppets used by the puppeteers and the order of their procession also reflects and reinforces the prejudices, fears, and desires of the society outside the cave.

The traditional philosophers of science have been heavily criticised for their neglect of experimentation.[5] There have been many studies of experimentation made within the last twenty years to highlight and counter this neglect. Writers from diverse academic disciplines have paid very close attention to how working scientists actually perform experiments, publish their findings, deal with controversy and criticism, and weave their work back into the wider culture from which it is emergent. It takes considerable effort and time to translate the experiences acquired within the local context of the experimenters' work into context-independent facts, general theories, hypotheses, or universal laws. There has been considerable focus upon the way that scientists negotiate the dissemination of their work from the local context of the laboratory to the wider public domain of shared representations, results, and facts.[6] Scientists have been studied at work and how they have chosen their materials, built, organised, and operated experiments, and produced representations, facts, and knowledge has been analysed at length. By paying close attention to the ways that social and material practices are innovated and ordered within the real-time processes of scientific work, the interaction between investigative, observational, and communicative techniques with the content of what is experienced or disclosed has been analysed in detail. The tactics that scientists use to reproduce findings and deal with the obstacles to communicating the techniques and knowledge needed to replicate new results have been examined at length. By examining the institutionalization of procedures for the evaluation and dissemination of expertise and knowledge, the social importance of material practices to the construction and communication of facts, concepts, techniques, has become central to the way that real experiments are understood within contemporary science studies. These studies bring the social and historical conditions of both scientific discourse and technological practices into question.

The purpose of this book is to reflect upon how the world is investigated, represented, and experienced by working physicists, whilst questioning how scientific knowledge of natural processes is possible on the basis of the use of artificial experimental apparatus, procedures, and techniques. The philosophical arguments presented in this book are premised on the supposition that we need to understand how the epistemology and ontology of modern physics have been historically related in the context of ongoing material practices. How does experimental physics achieve knowledge about the world? Or, to be more precise, what kind of knowledge, if any, do physicists achieve? And, if physicists achieve knowledge, which parts of the world do they achieve knowledge about? If experimental physics is an artificial process, an art, and the objects of scientific thought are artifacts, then what do we mean by artifice? How does theory relate to technology? How does technology relate to Nature? Physicists claim that they use technology for a very specific purpose of building and using machines to discover "the truth" about "the fundamental principles of the Universe" by working out how these machines work in terms of fundamental "natural mechanisms". Furthermore, they claim to be able to use their understanding of "natural mechanisms" to innovate new machines and, by doing so, increase their understanding of how invisible "natural laws" work. As well as building and using machines to acquire money, fame, glory, and for its own sake, as a pleasure, physicists also pursue their technical art to learn truths and satisfy their curiosity about change and permanence in the natural world. It is the use of machines to discover "natural mechanisms" and "the truth" that is the subject of this book. Do we understand machines? To what extent is the machine autonomous or controlled? To what extent is the machine operational in the realm of science and the way that we understand the world? Given that scientists turn to machines as an arbiter of fact, as non-human neutral instruments of measurement, then the machine is a co-participator in the construction of the judgement of what constitutes knowledge. How is this done? To what extent do machines create the facts of measurement and the reality that they disclose?

In order to answer these questions, I shall examine how experimenters experience and describe the things that they use to investigate the natural world. By examining how physicists experience the objects of scientific research, we can examine how they experience the way that these objects are presented as a mode of disclosure of a postulated invisible world of underlying reality. This involves examining the tacit presuppositions that are required to make those experiences and practices possible. Unmasking the presuppositions that are implicit in the history and trajectory of the theoretical and experimental practices of working physicists shows that there is frequently a gap between the practitioners' interpretation of their own practices and how those practices appear to one who does not share those presuppositions. Working practitioners unreflectively utilise tacit knowledge in their interpretations and actions. As Michael Polanyi pointed out, tacit knowledge is non-verbal and implicit to practices and is consequently not open to criticism by the practitioners because they are unaware of it.[7]

The reality that physicists claim to discover is a product of modes of agency that cannot be separated from the social and historical contexts of material practices and interpretations from which they sprang. It is the historical and cultural conditioning of the activities and practices of the experimenter, embodied through training, that readily permit a naturalization of the use and interpretation of instruments in the laboratory. It should come as little surprise that physicists are realists in the laboratory. As Robert Crease argued, when experiments are inherently a series of performances, interpretations are central to the act of producing measurements and observations from material events, whether or not scientists are aware of them.[8] These interpretations are only possible because science has a historically and culturally situated meaning which allows the judgements made by working experimenters in the real-time processes of laboratory experiments to provide meaning to their experiences and practices.

How are science and technology related?

It is often supposed by the traditional philosophers of science that theory precedes and anticipates experiment in the form of hypotheses, conjectures, or predictions. Experiments are simply designed to test them. They presuppose that science provides us with a rational understanding of Nature and modern technologies embody that understanding as "applied science". This "received wisdom" is that technology is the logical consequence of the application of scientific knowledge and rational thought to human problems in the material world. Technology is supposedly only a means to satisfy human purposes and has no role in shaping scientific knowledge, the conception of rationality, or human intentionality. This view presupposes that rational thought and logic transcend the material world and that the primary relationships between the human mind and the world are those of cognition, manipulation, and control. It presumes that technology enhances and extends the powers of the human mind and senses without changing or directing either. The construction of theories is a purely intellectual affair for which the technology of experimentation does not have any constitutive scientific role. The experimental apparatus and methodology are supposed to be ontologically and epistemologically neutral. Technology is ignored as largely irrelevant for the epistemology of science, as something that, at most, is a matter for applied ethics. In the traditional view of the "scientific revolution" of the sixteenth and seventeenth centuries the rise of modern technology is taken to be derivative from the mathematical sciences and modern technology is taken to be "applied science".[9] However, the proponents of this view have not provided us with a satisfactory account of how this "application" occurred. It was for this reason that Alfred North Whitehead considered the way that highly abstract mathematically formulated theories have been "effectively applied to practical affairs" to be the "paradox" of modern science.[10] How did the mathematical sciences lead to modern technology? How were mathematical and technological practices conceptually connected and justified? Ernst Nagel argued that the relations between modern science and technology are not as obvious and clear as the traditional philosophy of science has assumed.[11] He maintained the traditional view that modern technology is applied science but was aware that the character of "application" is an ambiguous one.

Traditional academic philosophy has been heavily criticised for its neglect of the centrality of technology to the human condition and the production of knowledge.[12] In contemporary philosophical studies of science and technology, considerable attention has been paid to the centrality of technology to human existence and knowledge to the extent that an "alternative tradition" has become fashionable. On this view, the experimental "natural" sciences, such as physics, chemistry, and genetics, are seen as forms of "applied technology".[13] Many historians of science and technology have also argued that technology preceded and led the sixteenth century "scientific revolution" and that modern science is "applied technology" to a lesser or greater degree.[14] It has become widely accepted that modern science is *technoscience* that can only be understood in relation to its uses within the culture in which it is emergent.[15] However, the characterization of a "natural science", such as physics, as "applied technology" only reverses the problem of how the "application" occurred. How was technology "applied" in such a way as to become the "natural science" of physics? In order to understand experimental physics as "applied technology" we need to address how mathematics, machines, and natural phenomena were related. What understanding of technology do we need in order to understand the technological basis of the experimental natural sciences? It is my view that we need to analyse physics at a "deeper" level than merely pointing out that the use of mathematics and technology has been central to the experimental natural sciences since their origin in the sixteenth century. In order to understand the conceptual possibility of experimental physics, we need to historically trace back its origin that permitted its current manifestation as a

modern technoscience. What presuppositions about both natural phenomena and technology permitted the use of technologies to understand natural phenomena? This is a question of the metaphysics that
underlies the whole legitimacy of the technological disclosure of
natural mechanisms. My argument in this book is that the metaphysics
of mechanical realism provided the operational precepts of experimental physics and made the epistemological use of technology to disclose
natural mechanisms and laws conceptually possible. How was this
metaphysics possible? When did it occur? My position is that this
metaphysics occurred during the fifteenth and sixteenth centuries
and, in fact, made the "scientific revolution", experimental physics,
and modern technology conceptually possible. As I shall argue in
chapter three, the precepts of mechanical realism allowed the mathematical description of the motions of the six simple machines (the
wedge, the lever, the balance, the inclined plane, the screw, and the
wheel) to be taken as descriptions of the fundamental natural motions.
We need to understand how the reification of mathematics, as something objectively, eternally, and universally true, in the context of the
Renaissance developments of the Medieval science of mechanics,
allowed experimental physics to be metaphysically operational as a
technological mode of disclosure of natural mechanisms, and for technology to be a consequence of the utilization of natural mechanisms in
material practices. This provided the sixteenth and seventeenth centuries with both a methodological and an ontological foundation for
the mechanical and experimental natural philosophies of Galileo,
Descartes, Bacon, Gassendi, Newton, Boyle, Hobbes, *et al*. In chapters
four and five, I shall describe how this methodological and ontological
foundation was central to the methodology, intelligibility, and subsequent researches of experimental physicists.

Martin Heidegger and Jacque Ellul offered insights into the nature of
modern technology and its relation to human existence.[16] Both were
concerned with the question of how we can attain a free relation
with modern technology. Ellul questioned the meaning of the dominance of technique for the human present and future, and Heidegger
was concerned with preparing a way in which we could question
the essence of technology and develop a free relationship with it. *The
Technological Society* is a narration of the tragedy of a civilization
increasingly dominated by technique, and *The Question Concerning
Technology* was an attempt to reveal the essence of technology and
relate it to truth. Ellul placed an emphasis upon the erosion of moral
values brought about by technicism, an examination of the role of

technique in modern society, and a historical disclosure of the forces that have shaped the development of technical civilization. For both Heidegger and Ellul, Western civilization is a progressively technical one committed to the quest for continually improved means to carelessly examined ends. What was once prized as a good in itself, for its own sake, is transformed into something that is only of instrumental value, for the achievement of something else. Whilst technology transforms ends into means, human beings are compelled to adapt to a technical substratum of human existence that has become so overwhelmingly immense that we are unable to cope with it as a means and, consequently, treat it as an end. Heidegger examined how the instrumentalist and anthropological definitions of modern technology, whilst being correct, have made us blind to the essence of technology. The anthropological definition is that technology is a human activity and the instrumentalist definition is that technology is a means to an end. For Heidegger, the essence of technology was not to be considered as something technological and he considered the claim that technology is "something neutral" to be the worst misconception of technology possible because it immediately delivers us over to an unthinking relation with it. He agreed that technology is a human activity in the sense of positing ends and procuring the means to them, and accepted that it is also an instrument in the sense that the manufacture and utilization of equipment, tools, and machines, as well as the needs and ends that they satisfy, all belong to what it is. However, if we are to understand the essence of technology, we need to ask: What is the instrument itself? Within what do means and ends belong?

For Heidegger, modern technology was a mode of disclosure in which beings are set in place, ordered, in such a way as to "put to nature the unreasonable demand that it supply energy that can be extracted or stored as such."[17] To use his examples, modern technology challenges a tract of land to yield coal and ore. The earth is disclosed as a coal-mining district and the soil as a mineral deposit. Air is set upon to yield nitrogen for the mechanised agricultural industry and the earth is set upon to yield uranium for the atomic weapons and power industries.[18] Modern technology sets upon and challenges Nature to disclose itself, unlock and expose itself, as energy or resources for future use. It is:

> "always itself directed from the beginning toward furthering something else, i.e. toward driving on to the maximum yield at the minimum expense. The coal that has been hauled out in some

mining district has not been supplied in order that it may simply be present somewhere or other. It is stockpiled; that is on call, ready to deliver the sun's warmth that is stored in it. The sun's warmth is challenged forth as heat, which in turn is ordered to deliver steam whose pressure turns the wheels that keep a factory running."[19]

It is this availability for use in the future (without any consideration of the particularity of that future use) that Heidegger termed as standing-reserve.[20] He used this term to characterise the way in which everything is commanded into place and ordered according to its utility for the challenging essence of modern technology as it comes into presence as a mode of disclosure. Objects lose their character as objects when they are disclosed as standing-reserve. It was for this reason that he considered the instrumental definition as something that was both correct and concealed the truth. Modern technology can only disclose Nature as standing-reserve because it challenges human beings to exploit Nature in this way. This challenging is an imperative in which the participation of human beings in the ordering disclosure is essential if it is to happen at all. By responding to the challenging, human beings are set-upon, gathered together, and ordered into modes of disclosure, and it is the way of disclosure that discloses objects as the *objectlessness* of standing-reserve.[21] Destining (*Geschick*) sets human beings on a way of disclosure of the real as standing-reserve and *Ge-stell* simultaneously sends and gathers human beings upon this way of disclosure. Heidegger rejected the idea that this involves "a fate that compels... where 'fate' means the inevitability of an unalterable course" and human beings do not control what is disclosed by this ordering.[22] Heidegger termed this way of disclosure as *Ge-stell*, which when translated as "Enframing" retains the connotations of "frame" and "skeleton". *Ge-stell* is not something technological (in the same way that pistons, rods, and chassis are technological) and the assembling of the technological, the ordering of the stockpile of components falls within technological activity. Technological activity is merely a response to the imperative, the destining of *Ge-stell*, and it neither comprises *Ge-stell* nor brings it about. Under the sway of *Ge-stell* everything is transformed into standing-reserve for future use. For Heidegger, it was precisely this monolithic character of *Ge-stell* that "threatens to sweep man away into ordering as the supposed single way of revealing, and so thrust man into the danger of the surrender of his free essence."[23] Thus, for Heidegger, the anthropological and instrumental definitions are correct but conceal the essence of modern technology.

Like Heidegger, Ellul considered technique to have its own reality, substance, and particular mode of being. Modern technology, or the technical phenomenon, was ontologically identical with the technical society. He characterised the technical phenomenon as a perpetual state of social inequilibrium and defined "technique" as "the totality of methods rationally arrived at and having absolute efficiency (for a given stage of development) in every field of human activity" and proposed that it should be studied as a sociological phenomenon.[24] "Technique" referred to any complex of standardised means (or ensembles of means) for attaining any deliberate, stable, and rationalised productive behaviour or intentions to achieve predetermined results. For Ellul, technique was objective in the sense that "it is transmitted like a physical thing" through the organization of productive performances.[25] Technique is the organised ensemble of practices that are used to secure any end whatsoever, and it can *in principle* only provide technical and quantitative solutions to technical problems. He rejected any metaphysical notion of "technological determinism" in his characterization of technique because things could always differ from the contingent actuality of the present. If technique is a "blind force" it is so because human beings have closed their eyes to the alternatives. Technique has the character of an imperative that only achieves its power because human beings respond to its demand. It is this conception of technique, in terms of a unitary imperative to order the world, which has a considerable parallel with Heidegger's conception of *Ge-stell*.

Ellul characterised the modern conception of "rationality" in terms of a "technical rationality" which brings mechanics to bear on all that is spontaneous. It operates through systematization, division of labour, creation of standards, production of norms, and the reduction of method. Any intelligible critical analysis of any productive technology, technical decision, or process, must be placed into a socio-economic context of political interests and organization. Every technology is a social ordering and organization of production processes directed to the satisfaction of socially (politically and economically) emergent goods (or ends). A technological order is a social order *and vice versa*; non-human artifacts are participants in the shape and direction of society. Technological choices reiteratively shape and limit the social, economic, and political landscapes.[26] "Technical rationality" is a context-dependent *bounded and evolving rationality*.[27] As Mueller put it

... the form taken by a technological system is not a *design* but an *evolutionary trajectory*. The trajectory is defined as its operations adjust to the specific social, geographical, economic and political characteristics of its environment, and overcome the technical problems posed by its growth and its competition with other technologies. Anyone familiar with the history of a large-scale technological system knows that an attempt to implement the simplest idea can create vast numbers of unforeseen problems. It is the process of solving these problems – not of preconceived design – that gives technology its shape.[28]

The horizon of possibilities and the ways to reach it are reiteratively shaped by the social, political, and economic choices and problems, for which technology has promised to solve. By bringing new things into the world, such as hydrogen bombs, antibiotics, contraceptives, radio, motorcars, etc., technology transforms the world. New political, economic, and social possibilities arise because new modes of social organization become possible. These possibilities, when realised, shape the directions of technology. Technological objects are non-linear complex objects that are emergent through interconnected socio-technical feedback relations and an evolving socio-technological background. It is no more neutral than the contexts in which it arises because the uses to which any technology is put to are the content of its evolution. Technical rationality is a bounded and evolving form of rationality directed in accordance with a bounded and evolving concept of "efficiency" as its basis for informed choice.

For Ellul, every intervention of technique is, in effect, a reduction of facts, forces, phenomena, means, and instruments to the logic of "efficiency". The human agent becomes transformed into an agent that is defined in terms of her/his performance and function, as an integrated and articulated component, in an ensemble of functioning agents. Technique sets upon and organizes human agency and is globalised through the educational and technical dissemination of values, projects, techniques, and technicians. It is a global unification of a monolithic mode of social organization and cannot be anything other than totalitarian because the imperative towards "efficient" intrinsically requires the absorption of any plurality of means into a "scientific" unification in order to maximise "efficiency", co-ordination, and exploitation.[29] This involves the total organization of the human population to achieve "efficiency" and "maximise results" in every area of human endeavour. It is this totalitarianism that generates the monopoly of a technical phenomenon that achieves its autonomy by being available as "the best technique". Technical civilization is entirely constructed in terms of technique to such an extent that only that which is technical is considered to be part of civilization. Everything must serve a technical end and anything non-technical is excluded as "inefficient", "subjective", or it is reduced to a technical form. The project of "the technical man" is a perpetual search for the "one best way" to achieve any designated objective and the perpetually expanding and irreversible role of technique is extended to all domains of life. The choice of method/technique is made in reference to the satisfactory stabilization of measurements, calculations, and productive practices, in relation to an intelligible causal account. Such a choice cannot be divorced from the socio-technical backgrounds against which it is made and emergent from. It is a matter of paradigmatic socio-technical consensus. Once this choice has been made then technique becomes a technological object available for future work and is placed in competition with other technological objects for ordering within the technical imperative towards "efficiency". The socio-technical winner is taken to be "the most efficient" and "the one best way". Until it is replaced by another technique, "the one best way" achieves a technical autonomy in practice because technical practitioners, also under the sway of the technical imperative, are obliged to use it. Its results are indisputable until a "better" technique takes its place. Once a technique has become established as "the one best way" then it is no longer an object for technical deliberation. The technical practitioner is committed, whilst under the sway of the technological imperative, to perform her/his operations in "the most efficient" manner which, of course, requires using "the one best way".

Ellul was concerned with the way that the dominance of the technological imperative has taken over all human activities and ordered everything according to its utility according to mechanistic logic, techniques, and machine processes. Like Heidegger, Ellul warned us against considering the essence of technology as something technological. The all embracing technique was, for Ellul, the "consciousness of the mechanised world" and it is this "consciousness" which manifests itself as the imperative to integrate everything into the mechanized world and "will assimilate everything to the machine; the ideal for which technique strives is the mechanization of everything it encounters."[30] Technique has become autonomous and integrated substance of human agency within society. It led to the mechanization of human activities, via the action of the machine, by applying the "know how" of mechanization to domains which were previously "lacking" machines. The technical imperative to find and use the "most efficient" means pervades all human activity and "ranges from the act of shaving to the act of organising the landing in Normandy, or to cremating thousands of deportees."[31] Every material technique is subordinate to its immediate result and "efficiency" is determined by choosing the technique that produces the most satisfactory result. The technical phenomenon is artificial, lacking spontaneity, and opposed to Nature. The technical imperative is an attempt to create an artificial system that is supposedly more intelligible, controllable, and conducive to human well being than the natural world. It is this attempt that Ellul considered to be "the societal gamble": a gamble on the superiority of an artificial world over the natural world. It is the whole technical phenomenon, the technological society, which is itself a gamble. The artificiality of the technological society created through technique destroys and replaces the natural world. It does not even allow the natural world to restore itself or enter into a symbolic relationship with it, and, accordingly, these two worlds are incommensurable. Just as hydroelectric installations take waterfalls and lead them into conduits, so the technical milieu absorbs the natural. Heidegger also used the example of the hydroelectric plant to describe the way that the Rhine is disclosed as hydraulic pressure for an interlocking complex of turbines, electromagnets, power stations, and a network of cables, setup to provide electricity as standing-reserve.[32] The river is damned up into the power plant and is transformed into a power supply. Even to the extent that it is still a river in the landscape, it only remains so as an object for the tourist industry. We are rapidly approaching a time when there will no longer be any natural environment at all.

How did Ellul and Heidegger relate science and technology? For both Ellul and Heidegger, mathematical techniques were central to their definitions of science and they considered only that which can be expressed numerically to be scientific and, hence, the scientific use of technique is that of reducing the possibilities of investigation to the calculation of numbers. Although Heidegger accepted that modern physics "as experimental, is dependent upon technical apparatus and progress in the building of apparatus", he argued that human beings are challenged and destined by *Ge-stell* to disclose Nature as the standing-reserve of energy, and this attitude, on the part of human beings, was first displayed in the rise of modern physics as an exact science.[33] Physics was a "way of representing [that] pursues and entraps nature as a calculable coherence of forces" and even as pure theory "sets nature up to exhibit itself as a coherence of forces calculable in advance, it therefore orders its experiments precisely for the purpose of asking whether and how nature reports itself when set up in this way."[34] He maintained his view that "mathematical physics arose almost two centuries before technology" but claimed that because "physical theory prepares the way first not simply for technology but for the essence of modern technology" then modern physics "is the herald of Enframing, a herald whose origin is still unknown."[35] He argued that, despite the fact that "chronologically speaking" modern physics began in the seventeenth century and machine-powered technology began in the latter part of the eighteenth century, the essence of modern technology was "the historically earlier". Modern physics was itself challenged forth by *Ge-stell* in the demand that Nature is orderable as standing-reserve. This set-up physics as a means by which Nature was disclosed "in some way or other that is identifiable *through* calculation and that it remains orderable as a system of information."[36] Thus for Heidegger, it is for that reason that "modern technology must employ exact physical science".

Ellul accepted that it is the creation of general explanatory theories that makes science distinct from technique. However, he was critical of the view that modern science is pure theory and technology is applied science. He argued that this view is "radically false" because it is only true of the nineteenth century physical sciences and is not true of science and technology in general.[37] In his view, technique preceded science ("even primitive man was acquainted with certain techniques") but only began to develop and extend itself when science appeared. Technique required science to progress because it had to wait for science to provide the solutions to the problems posed by the repeated experiments of technique. How did science provide the solutions? What were the problems? Ellul did not address these questions. He merely maintained that the border between technological and scientific activities is not sharply defined and that technique provides preparatory work for scientific synthesis. Science has become the instrument of technique because scientific discoveries are increasingly implemented in every day life before the consequences of that implementation have been considered. For Ellul, modern science became bound up with technique during the Industrial Revolution and development of the machine and the application of technique to all spheres of life. The work of technique, the mechanization of all human spheres of action, was a systematization, unification, and clarification of everything. Modern scientific research increasingly requires large teams of researchers, enormous amounts of money, and the aid of machines. The work of large scientific research is increasingly technical work and "pure science" is becoming increasingly "applied technology". Ellul observed that science has been becoming increasingly governed by technique since the nineteenth century to such an extent that the smashing of the atom and the smashing of Hiroshima (and Nagasaki) are manifestations of the same imperative.[38] Science requires the application of technique as a necessary condition of its existence and without technique science is merely hypothesis and theory. He also used the example of the way that the steam engine was the product of technical trial and error sequences of invention and improvements and scientific explanations came much later, to illustrate this point. However, he did not provide us with an account of how those explanations were forthcoming. Nor did he show us how they were related to invention. How did precision and explanation solve the problems of technique? Ellul did not explain this to any greater depth than arguing that there is an increasing interaction between scientific research and technical preparation to such an extent that science is incapable of progressing without the technical means to do so. For example, according to Ellul, Faraday was unable to precisely formulate his theories about the constitution of matter because of a lack of high-vacuum techniques. But why should this be the case? If technique and Nature are independent, as Ellul maintained, then why is technique necessary for natural science? Did he mean to imply that somehow high-vacuum techniques were necessary for the precise formulation of Faraday's theories? Surely, if technique does not have anything at all to do with Nature then there would be not any necessity for any particular technique for the scientific synthesis to proceed. What was the "matter" that Faraday wished to explain? Ellul did not address these questions either.

There was considerable equivocation and inconsistency, on Ellul's part, in his description of the relations between science and technology. He maintained the view that mechanical progress "is limited by the physical world" and that the drive for efficiency is the mobilization of "the forces of nature" and an "intervention into the inorganic world."[39] However, he also insisted upon the fact that the "only thing that matters technically is yield, production. This is the law of technique; this yield can only be obtained by the total mobilization of human beings, body and soul, and this implies the exploitation of all human psychic forces" and, yet, the "new milieu has its own specific laws which are not the laws of organic or inorganic matter... Man is still ignorant of these laws."[40] Does this equivocation reveal an inherent contradiction in Ellul's argument? How could science and technology be related in this way and have nothing in common with the natural world? Did Ellul take "the physical world" and "the inorganic world" to be distinct from "the natural world"? Unfortunately, he did not discuss the relations between these "worlds" further. In my view, the equivocation in Ellul's thesis was a consequence of his rationalist account of mathematics and characterization of science as mathematical and explanatory, on one side, and the absence of any account of the technique(s) which provided a clear link between mathematics and technology, on the other. What was the scientific synthesis? How did science produce explanations? Ellul argued that the precision of any machine is only possible because of the elaboration of its design with mathematical rigor in accordance with its use, but he did not provide any account of how this elaboration of its design could be performed. This meant that practical activity rejected gratuitous aesthetic preoccupations in favour of the idea that the line best adapted to use is the most beautiful. It was supposedly necessity and the certainty of mathematical calculation that characterised the technical world but he did not provide any account of how mathematics became bound up with the technical imperative and conceptions of "efficiency". He accepted that this imperative somehow developed out of the science of mechanics but its origin was "mysterious and enigmatic" and possibly bound up with magical rituals.[41] This omission lead to the considerable equivocation on his part that can be seen in his stance that modern science depends upon technique whilst maintaining the traditional view that *somehow* technology is "applied science" because it is mathematical.

My argument in chapter three is that if we take a closer look at the transformation in the status of mechanics and its internal construction in the sixteenth century then we can see how the confidence in and the value of the practical advances of the technical imperative was present in European thinking from at least the sixteenth century onwards. In fact, we can see a "clear technical intention" in the writings of Francis Bacon, the practical interests of the Royal Society, and the efforts of the sixteenth century Italian mechanists and engineers. The practical values of the new sciences were central to the whole enterprise of the new sciences from their very beginnings. Craft practices and the innovation of novel tools and instruments were central to the work of Galileo, Descartes, and Newton. The technical imperative was present in the sixteenth century drive for the achievement of commercial, political, and military advantage in the competitive contexts of European ambitions. It may well have been a dream but the intention was there. The destining of modern technology had begun. Furthermore, individualism was apparent in the natural and political philosophers throughout the seventeenth century and eighteenth century. The identification of "Man" as an isolated and rational individual is apparent in the writings of Descartes, Bacon, Bentham, Hobbes, and Rousseau. The focus upon material practices and social relations based upon their satisfaction of individual interests preceded the nineteenth century. This individualism was itself a consequence of the beliefs in the primacy of the techniques of rational and reasoned discourse and the unitary material relation between "Man" and "Nature" as disclosed by the mechanical philosophies.[42] These conceptions were a consequence of the technical imperative rather than its conditions and were the conditions for the Industrial Revolution and the mass participation in the technical imperative to be represented as progress, social evolution, and human destiny.

Ellul located the search for "efficiency", the demand for the "one best way to do work", in the nineteenth century. However, it was the mechanical philosophies of the sixteenth and seventeenth centuries that proposed that there is one single most efficient mechanism in operation between any particular cause and its effect(s). That "most efficient mechanism" was termed as "the natural mechanism" and it was the allotted task of the natural experimental philosophies to find it for any particular cause-effect sequence. The metaphysics of mechanical realism provided the link between the mathematical sciences and the practical sciences. The distinction between "pure" and "applied" science is *merely* the distinction between finding the "most efficient mechanism" and implementing it in the organization of productive practices. Once we address the extent that experimental physics interactively and reiteratively involves both the discovery and implementation of "the most efficient mechanism" in ongoing research and technological practices, we can characterise experimental physics as internally both "pure" and "applied". In my view, this characterization is as appropriate for the mechanical philosophies of the seventeenth century as it is for the physical sciences of the nineteenth century (and for the twentieth century as well) because both shared the same metaphysical precepts. It is merely the case that it is more obviously characteristic of the nineteenth century physical sciences. It was the project of applying Euclidean geometry to the problems of mathematically describing the six simple machines that linked mathematics and mechanics. The metaphysical precepts of mechanical realism allowed these mathematical descriptions to be presented as representations of "natural laws" and opened the way for the possibility of using mathematically described mechanisms to explain the occurrence of natural phenomena. This metaphysics allowed technique to be understood as limited by "the physical world" and as an intervention into "the inorganic world" by mobilising "the forces of Nature". However, as I shall endeavour to argue throughout this book, these "worlds" are nothing more than the abstractions of the whole complex of technical ensembles of machines, in which any mobilization of "forces of Nature" is the non-linear interactions that occur during the attempts to integrate novel machines into this complex. It does not necessarily have anything to do with the natural world at all! This complex is itself only a small part of the real world (which I take to contain both natural and artificial entities). If we are to understand the origins of physics as a technoscience, we must inquire into how the massive reduction of the ontology of the real world to an innovated collection and ensemble of machines was metaphysically founded. This involves an inquiry into the relationships between technology, knowledge, and truth.

Technology, knowledge, and truth

Heidegger was critical of the way that "know-how" (*techne*) is treated as the ultimate virtue in the modern age.[43] Heidegger based his distinction between *Ge-stell* and *techne* upon the change in the destining of disclosure. For Heidegger, the ancient handicrafts were a different mode of disclosure from modern technology because they participated in "bringing-forth" (*poiesis*) beings into the world as ends-in-themselves. They were intimately bound-up with *aletheia* (truth) as a mode of disclosure and presencing of the real. This truth was bound up with modes of completion and perfection and did not correlate with the definition of truth as "correctness" (*veritas*).[44] *Techne* (plural: *technai*) had a loose meaning of art, craft, or science in pre-socratic Greek.[45] It had the connotation of "device" in the straightforward sense of "ploy" rather than "something devised". It had similar meaning to "crafty" and "artful". In ordinary usage *techne* was used to refer to cleverness and cunning in getting, making, or doing, as well as to trades, crafts, and skills of every kind. It involved a collection of tactics, stratagems, and tacit "know-how" as kinds of activity to achieve specific ends. It was in the philosophical writings of Plato and Aristotle that *techne* was treated as a formal kind of knowledge, which could be used as a theoretical guide to govern making. For both Plato and Aristotle *techne* referred to the general, abstract, and communicable first principles of making and inscription in the activities of craftsmanship and art. It is how *techne* was related to *episteme* (commonly translated as "science" or "knowledge of eternal and necessary principles") that differed between Plato and Aristotle.[46]

In Plato's works, *techne* and *episteme* were closely related when discussing art and knowledge in general and were used interchangeably to characterise geometrical reasoning in particular. In *Philebus* (55c–56d), *Statesman* (258e), *Gorgias* (450b–c), and *Ion* (532), for example, Plato used *episteme* to describe mathematical truth as eternal and necessary knowledge, whilst using the word *techne* to describe mathematics (including logic, arithmetic, and geometry) as the highest form of art. Socrates often argued that all *technai* are involved with *logoi* (words, speech, reason, principles) bearing upon some specific subject matter of the art in question, even though some require a great deal of physical exertion and very little reason (i.e. horse riding, painting, or sculpture) and others require a great deal of reason and very little exertion (i.e. arithmetic, logic, or astronomy). Only routine activities (i.e. cooking or persuading), unreflectively based upon experience and habit, devoid of *logoi* were considered to be *atechnos* (devoid of art). Of course such activities, such as cooking or persuading, can be (and are) developed into arts, but for everyday purposes the usually are not. According to Socrates, what such activities lacked, in order to qualify as *techne*, was knowledge of the *aitai* (intelligible causes) involved in what was made or done. Such everyday habitual and unreflective practices were *alogos* (without words, reason, or principles). Any productive activity (*poiesis*) needed to be teachable through *logos* in order to qualify as *techne* and henceforth *techne* was the knowledge of all productive activities that could be reasoned about and taught. It was the logical and communicable knowledge regarding the causal principles involved in making or doing something. It could either proceed by conjecture and intuition based on practice, training, and instruction (i.e. music, medicine, or agriculture), or it proceeded through the use of numbering, measuring, or weighing. The mathematical activities of numbering, measuring, or weighing were taken to be the most truly *technai* because they were taken to involve the greatest precision and were more closely associated with the activities of making that operate upon the material world. These reasoned activities operated by guiding acts of making through the use of mathematics, and the *techne* of such activities, provided a formal knowledge and rules by which material practices were performed, governed, and understood. However, in

Philebus (56d), *epistemoi* such as arithmetic were distinguished from *technai* such as carpentry because the former deals with abstract numbers whereas the latter uses numbers to deal with materials. In *The Statesman* (258e) *episteme* was used to denote pure theory or any knowledge that did not relate to the material world in a practical manner. *Episteme* was reserved for knowledge learnt for its own sake.

Aristotle, following Plato, also defined *techne* as a kind of knowledge of making or production that informed material practices. He used the word *techne* to refer to any theoretical knowledge concerned with making that was explanatory, generalised, abstract, formal, and communicable (*NE* 6.4; *Metaphysics* 1.1; *Rhetoric* 1.2). *Techne* was induced from unarticulated particular experiences and practices into communicable, formal, and general knowledge of the first principles (or intelligible causes) involved in making or producing something. It was to be used to reason about how to make particular things in a specific manner. It was the general knowledge of the principles and causes, the know-how and the know-why, of any specific art or craft. It was inextricably bound up with an intellectual grasp (consciousness or cognition) of first causes that provided the kind of knowledge possessed by an expert (*technite*) in any one of the specialised crafts. *Techne* provided "a true course of reasoning" that guided stable dispositions to make particular things or bring about a state of affairs in a specific manner (*NE* 6.4.1140a11). It was distinct from experience because the latter could only be related to the particular, whereas the former was concerned with the general and was to be used to explain the particular. Someone may have the experience that outcome B will sequentially follow action A but, without a complete account of why B follows A, that person would not possess *techne*. *Praxis* (habitual practices) could be learnt from experience and mimicry, and be used to develop tacit, non-verbalised skills and beliefs regarding the best way to proceed. However, it was only when this acquisition of experiences and instruction had been completed (inductively abstracted in a general true course of reasoning) that *techne* could be acquired. The craftsman needed to give a "rational account" of *praxis* before s/he could be said to be a *technite*. This "rational account" was to facilitate the tracing back of a product to its causes.

Every *technai* has its appropriate forms, tools, and materials. These materials, tools, and forms govern and guide *poiesis* (*Metaphysics* 7.9.1034a10–11). Aristotle argued that the materials used in production were distinct from the *technite* and *techne* was not contained in the produced thing or state of affairs. Aristotle made a distinction between things that find their origin in the maker (*poieta*) and things that find their origin in themselves (*phusika*). The activities of bringing-forth (*poiesis*) were taken to bring about and terminate in a product, outcome, or end (*telos*) that was separate from them. A pot is brought-forth through the actions of the potter, whereas a tree is brought-forth in accordance with an internal principle of change (*phusis*). Aristotle considered *poiesis* guided by *techne* to be distinct from *phusis*, yet he used his conception of *techne* as his primary analogy in his elucidation of his conception of *phusis*, whilst maintaining the autonomy of the latter (*Physics* Bk.2, especially 2.2.194a22ff and 2.8.199a15ff). He used *techne* to elucidate his conception of *phusis* as teleological (frequently requiring *tuche*, meaning luck or chance, as a tripartite division). When *telos* was introduced through the activity of a *technite*, the source of change was separate from the thing in which the change happens. Something could only be considered to be *phusika* when the source of change was immanent within the thing itself. *Techne* was the possession of the most helpless, unshod, unarmed, unclad, but highest animal who could, through *techne*, turn this weakness around, take advantage of *phusis*, and even complete that which *phusis* left incomplete (*Physics* 2.8; *Politics* 1337a1–2). For Aristotle, *techne* was rooted in and a completion of *phusis* to the extent that even human nature was completed by *techne* through medicine, crafts, and politics (*Physics* 193b10 and 2.1.193a12–17; *Politics* 1.2.1253a2). Art imitates and completes Nature by attempting a union of form and matter that achieves a deep union in which the *telos* (the end) comes from within. Thus, for Aristotle, *techne* was directed towards perfection and the *technite* must attend to the materials s/he works with. Within Aristotle's four fold causality of formal, material, final, and efficient causes, it was the *technite* who took on the role of efficient cause. As "a true course of reasoning" *techne* was taken to be contained within "the soul of the craftsman" as "a reasoned state or capacity to make" and was consequently taken to be bound up with the maker. It guided the hands to perform definite motions that moved the tools and shape the materials into the product. This motion embodied form (*eidos*) into matter (*hyle*) and produced substance or informed matter. *Hyle* should not be confused with the post-sixteenth century conception of matter as inanimate and structured material.[47] *Hyle* referred to an unknowable and incognate formlessness, the formless potential to receive form that is active in the reception of it (*Metaphysics* 7.9.1034a10–11). For Aristotle, no two lumps of clay were alike. *Hyle* was the particularity of any particular lump of clay and did not refer to the clay-like properties of the substance called "clay". It referred to the particularity of the particular (*Metaphysics* 7.8.1033b20–1034a7). It referred to the way that a potter is unable to make the same pot twice and the way that each and every pot, as well as the experience of making them, is different even though they are all made out of the same substance. It was this active and emergent particularity to which *hyle* referred (*Metaphysics* 7.8.1033b20–1034a7). Form could not be forced upon (or into) *hyle* because of this active character in the reception of form. The *technite* had to be responsive to the way that *hyle* received form and, although the form was in the soul of the craftsman, its union with *hyle* was directed by both *techne* and *hyle*. The extent to which the *technite* could impose form upon *hyle* was not entirely within the control of the *technite* and there was a definite limitation to the extent that *techne* could guide this union.

Techne guides the inscription of form into *hyle*, which is the formless potential to receive form that is active in the reception of it. There is a definite limitation to the extent that *techne* can guide the imposition of form in *hyle*; it is only to the extent that the intervention can be grasped by "the rational part of the soul", as form (*eidos*), that it can be known and a part of *techne* (*Physics* 2.2.194a23). Although the form is in the mind of the craftsman, its union with *hyle* is partially directed by *techne* and partially directed by *hyle*, but the craftsman does not control extent to which *hyle* can be informed. S/he must attend to *hyle* and be responsive to the way that *hyle* actively receives form. It was for this reason that Aristotle argued that both *techne* and perception (especially touch) were required to guide the activities of making (*NE* 2.9.1109b23). The *technite* had to be responsive to the receptivity, capacities, tendencies of *hyle* emergent as the particularities of the materials during attempts to impose form upon them, just as much as s/he needed to know the appropriate forms, tools, materials, and how to combine them. The receptivity, capacities, and tendencies of *hyle* emergent during *poiesis* would not have occurred without the intervention of the *technite*, but *hyle*, as the particularity of the particular, resists the imposition of the generality of form from having complete sway. Due to *hyle*, individual experiences of making could not be known in their particularity through the general *logos* of *techne*. Due to the generality of *techne*, the particularity of individual experiences could be emergent *qua* particularities. No general principle is capable of being applicable to all particulars (*N.E.* 5.10.1137b13–15). It was *hyle* that resisted the characterization of any *praxis* or *poiesis* under a single set of rules (or instructions) that could be communicated from *technite* to apprentice. Although *techne* was comprised of formal, communicable, general, and abstract principles of making, it was primarily learnt through imitation and attending to the particularity of the appropriate materials. According to Heidegger, the term "appropriateness" also gives meaning to the term *hyle*.[48] In ordinary Greek *hyle* meant forest, thicket, or woods, in the sense of a place for hunting and gathering material for building. From this ordinary meaning, *hyle* came to mean material for any and every kind of building or production. However, *hyle* did not mean raw material. It meant the capacity, or the appropriateness, for use in the construction of a product. The wood to make a table is selected and cut to order and, consequently, the very character of its appropriateness is decided in relation to the making of the table. It is in this sense that the properties of a natural entity, say a tree, are determined in relation to its appropriateness for a task of making when those properties are determined through planned interventions. Theory is an incomplete guide to action and human beings become builders by building (*NE* 2.1.1103a35). Aristotle defined *techne* in terms of knowledge of the changeable and temporal, whereas *episteme* was reserved for knowledge of the eternal and unchanging. *Techne* was a general knowledge of the Being of Becoming, whereas the particularity of activities evaded complete capture by generalities. Thus, for Aristotle, *poiesis* guided by *techne* was straddled on a continuum between particularity of practice and the generality of theory (*Metaphysics* 1.1.980b25ff).

How do these ancient conceptions of knowledge relate to a modern science such as experimental physics? If *technai* are involved, which definitions of *techne* are appropriate at which stage of the process of constructing experiments? Given the fact that *techne* is the term for craft-knowledge, and the Ancient Greeks are not famous for their experimental practices, it may well seem odd to the reader to characterise the knowledge at work in highly technological modern experimental physics in terms of ancient craft knowledge.[49] My argument is that, in order to understand the role of *techne* in modern experimental physics, we need to also understand the role of craft practices, mathematical practices, and *hyle* in experiments. The pre-socratic usage of *techne* captures something of Ian Hacking's use of the term "intervention" and, as I shall discuss in chapter six, the word *hyle* captures something of Andrew Pickering's use of the term "material agency" to describe the emergence of resistances during productive practices, and it also captures something of David Gooding's use of the terms "the participation of Nature", "recalcitrance", "phenomenal chaos", and "plasticity", in his description of the development of craft practices in the early experiments by Michael Faraday. Plato and Aristotle's definitions of *techne* were premised upon knowledge providing the highest degree of communicability, precision, and repeatability, on the basis of "a true course of reasoning". This "true course of reasoning" is given in terms of the unchanging causal principles of change and, as the knowledge of the Being of Becoming, is highly characteristic of the theoretical knowledge of modern physics. This sense of the word *techne* is an important one for the characterization of scientific knowledge aspired towards during experimentation because "the true course of reasoning", in modern physics, involves the reduction of natural processes by the question "how does it work?" It is constructed in terms of the mathematical representations of natural mechanisms and causes as guides to human interventions in productive and experimental labour processes. As I shall argue in chapter three, since Moletti and Galileo, the conflation of *techne* and *episteme*, based on the postulation of mechanical realism and the universality and eternality of the mathematical science of mechanics, has provided experimental physics with an epistemological warrant for accessing the ontology of Nature by inscribing machines and mechanical ensembles in terms of causal mechanisms. It is this sense of *techne*, as an ideal, that is applicable to experimental physics. If we take *techne* to be characteristic of the ideal knowledge of causal accounts of the processes of building repeatable experiments, then *hyle* provides us with a meaningful term to express part of the phenomenological experience of experimentalists of the way that experiments "do their own thing" and resist perfect reproduction. *Techne* is concerned with complete knowledge given in terms of "a true course of reasoning" and such knowledge, should it occur in experimental physics, would be the *end result* of experimentation. It promises to provide the abstract, general, and communicable knowledge of how to repeat the experiment.

In chapter five I shall argue that experimental physics is an art aiming to achieve its own *techne* for its own sake as well as a manifestation of *Ge-stell* destined to perpetually use and test the truth of itself by gathering and ordering itself as standing-reserve for future innovation and experiment. It is also situated within a wider world that provides resources in exchange for novel prototypes. The technological objects produced in physics are simultaneously produced for their own sake and for the instrumental value they have in satisfying the challenges of ongoing research and the wider world. The imperative of *Ge-stell* is the transcendental precondition of modern technology that gathers together human beings, challenges us (or sets us up), to reveal reality in abstract and non-sensuous truths that are tested in terms of their availability for future use. It is a technological attitude towards truth and the world that sets upon human beings and challenges us to set upon and challenge truth and the world through making change happen. Modern science cannot be understood simply in terms of "applied technology" because, through the use of mathematics, it represents itself as directed towards the realization and acquisition of *episteme*, and it thereby conceals the extent that it is a metaphysical art directed towards the technological acquisition of *techne* of Nature. Modern science, directed towards the acquisition of the *techne*, is enframing and directing the trajectory of the human understanding of Nature, whilst simultaneously representing *episteme* as something technological. Modern physics is a *destining* of human beings towards the metaphysical realization and acquisition of the *techne* of Nature and the new powers that it promises to bring forth. It is through the deepening of this destining in the science of cybernetics, in terms of its transparency through embodiment and the illusion of the "steersman" metaphor, which brings with it an increasingly unquestioning relationship with modern technology.[50] This brings with destining a danger to our chances of developing a free relation with technology. It also has the most profound impact upon our understanding of the world.

Heidegger in *Being and Time* addressed the question of science and technology existentially, emphasising the primacy of practical over theoretical concerns, and questioned the scientific objectification of Nature.[51] He was deeply concerned about the Galilean and Cartesian subject-object distinction that presupposes a subordination of the practicality of knowledge in favour of a relentless search for abstract theoretical knowledge. Heidegger's characterization of modern science was that modern science is an objectification of Nature that represents "it" in mathematical terms that cannot account for the "earthiness" of the world. Hence, modern science, as a theoretical technology, allowed the possibility of producing objects without true individuality (or thinghood). For Heidegger, the essence of materialism was concealed by modern science because it not only asserts that the world is solely comprised of the physical interactions between particles of inanimate matter, but also because it presupposes a metaphysics that reduces every being to material for labour. In the 1949 version of his *Letter on Humanism*, he noted that modern science had become the new metaphysics into which philosophy was becoming dissolved.[52] The unity of this metaphysics was unfolding, in a new way, in the science of cybernetics and, as a consequence of this metaphysics, the power of modern science belonged to *Ge-stell*. It could not be stopped because *Ge-stell* obscured the place of "the event of appropriation" (the origin of modern science). This was itself possible because of the characterization of knowledge as mathematical projection, the characterization of scientific investigation as experimental research, and the characterization of science as an ongoing activity. Heidegger argued that modern science treats *phusis* as if it were a self-making artifact and has been interpreted as if it were a kind of *techne*.[53] He posited that this interpretation of *phusis* is a consequence of the modern metaphysical conception of the essence of Nature as a "technique". Thus modern technology as *Ge-stell* and *techne* as a mode of *aletheia* were bound together with truth, and, as a form of truth, modern technology is grounded in the history of metaphysics.

As I shall argue in chapter three, Heidegger's characterization of modern science as being founded upon the metaphysics of mathematical projection offers us a profound insight into the character of modern physics and technology. Heidegger explored how the object of modern metaphysical reflection was determined in relation to a decision regarding *what is* and *the essence of truth*. What ontological interpretation of truth provided the basis for the foundation of modern science and technology? Under which metaphysics was the essence of modern science and the essence of modern technology brought together and connected? Heidegger considered the interpretation of modern technology "as the mere application of modern mathematical physical science to *praxis*" to be a misinterpretation.[54] Why is this interpretation a "misinterpretation" and how did it arise? I agree with Heidegger that machine technology, as the most visible outgrowth of the essence of modern technology, was not simply "the application" of modern science, but, as an autonomous transformation of *praxis*, it made demands upon and shaped the form and trajectory of modern science. However, Heidegger did not describe how this happened and considered its origin to be mysterious.[55] Thus he was unclear about how and why it was possible and, consequently, in my view, he equivocated on the connection between modern science and technology. Furthermore, he was unclear about how the content of the projected ground plan of scientific research was refined and corrected as an ongoing activity. Heidegger was correct to characterise both modern science and technology as being bound together, but he did not explain how they were bound together. How was this possible? How does the power of modern science belong to *Ge-stell*? How are modern science and modern technology metaphysically connected? Heidegger did not provide us with satisfactory answers to these questions. However, as I shall argue throughout this book, a closer analysis of how and why physics is performed reveals that the direction of research in experimental physics is driven by an imperative towards the novel and productive disclosure and implementation of mechanisms in novel kinds of machines. It is the disclosure of these mechanisms that binds truth with productivity, and metaphysically connects modern science and technology with knowledge. Heidegger's theoretical preconceptions about the objectification of Nature within physics became an obstacle to a deeper inquiry into its metaphysical foundation, and (despite being highly critical of positivists) he had a positivistic conception of the object of experimental inquiry as being that which, in all but the most recent phases of modern physics, was directly accesible to perception. As a consequence, he proposed a bipartite ontology of law-phenomenon and was unable to reveal the connection between modern technology and physics. As I shall argue, the ontology of experimental physics is a tri-partite ontology of law-mechanism-phenomenon. The objective of experimental physics has always been to disclose the mechanism that connects phenomenal changes with the law that supposedly governs those changes. The abstraction of mathematical laws is only possible after this work has been done.

Labour processes are social processes in which they are organised upon the positing of ends. Georg Lukács termed this positing as "teleological positing".[56] As I shall argue in this book, the labour processes of experimental physics, like all labour processes, have a teleological character. They aim to satisfy the purposes and challenges, which are central to the setting-up of any experiment, and they are situated within the technological framework of the research programme. The practices and choices adopted in the execution of any experiment are made with the explicit aim of satisfying those purposes and challenges, whilst contributing to the research programme. Genuine novelty in experimental physics is the product of a technical complex of heterogeneous component inventions combined into an ensemble, which is integrated with other ensembles in order to reproduce functionality. All technological objects are complex and cannot be understood without addressing their connections and interactions with other technological objects and the purposes that they were constructed to satisfy. Converging and integrating their components into a stable and unitary centre of transformative power produces all novel technological objects. The labour processes of experimentation are the refining processes of stabilization directed towards producing transformative powers. When labour is performed for its own sake then the labour process is an art directed towards its own self-perfection. All arts are self-directed labour processes that produce themselves for their own sake. Science participates in discovery by discovering its own possibilities. The purpose of the art of experimentation is to explore its own possibilities by making them happen. It is metaphysical in the sense that it aims to equate and unify these interventions and material practices under an operational principle derived from non-empirical conceptions of science, technology, and natural phenomena. By showing how these metaphysical precepts are related to modern technology, through concepts of "mechanism" and "law", it is my intention to show how experimental physics has been successful in the discovery of novel phenomena and achievement of predictive accuracy, whilst maintaining a critical distance from the scientific realists' application of the concepts of "natural mechanism" and "natural law" to natural phenomena. Hence it is my intention to show that the existence and progress of experimental physics does not support or refute the mechanical realist metaphysics that lies at its epistemological heart, and, consequently, the technological successes of physics cannot be used by scientific realists to support their claims about the ontology of the world. However, far from supporting a positivistic interpretation of physics, it is questionable whether a natural science based upon a metaphysical project of experimentation using machines and instruments is an empirical science at all!

2
The Spirit of the Enterprise

In this chapter I shall describe how experimental physics is an inherently realist enterprise to discover the "natural mechanisms" and "natural laws" that are presupposed to pre-exist the attempt to discover them. To this end, I shall describe in detail the operational metaphysics of mechanical realism that makes experimental physics conceptually possible as a means to obtain knowledge about the structures, mechanisms, laws, and content of Nature. This metaphysics provides a unifying conception of "the physical" that underwrites the foundational principles and assumptions justifying the whole epistemological enterprise of experimental physics as a natural science. It is premised upon the following set of operational precepts that provide principles of action and function as a technical guide for the conceptual establishment of a methodology to explore Nature:

(i)   Natural and technological phenomena both share a unitary origin;
(ii)   Both natural processes and machine performances come into being by the same causal principles;
(iii) There is a unique, eternal, and universal cause for every effect (or set of effects);
(iv)   The connections between causes and effects are the fundamental mechanisms of Nature;
(v)   The realization of any mechanism is governed by a Natural Law and, consequently, the performance of any machine is governed by Natural Law;
(vi)   The mathematical descriptions of the motions of mechanical devices, and machines, are mathematical descriptions of Natural Law; and,

(vii) The only distinction between natural phenomena and machine performances is that the latter require human intervention to come into being whilst the former do not.

This kind of metaphysics is distinct from speculative metaphysics, such as Gassendi's seventeenth century atomic theory of matter, or Newtonian absolute space and time, or Everett's many-world interpretation of quantum mechanics.[1] These were proposed in order to make particular experiments and observations intelligible in terms of particular conceptions and interpretations of Nature. However, they are replaceable and contingent within experimental physics, whereas the whole project of using machines to explore natural mechanisms and laws universally presupposes the operational metaphysics of mechanical realism. It is a requirement for experimentation using machines to be a means of ontological access and epistemological warrant, and has been implicitly presupposed to interpret the results of experimental physics since Galileo to the present day. It has endured through the subsequent paradigm shifts from mechanical physics to quantum mechanical physics, and from Newton's classical physics to Einstein's relativistic physics.

Any understanding of experimental physics should be based upon an understanding of technology, when experiments use technologies as the means to explore the natural world. Technology is not a neutral means if the directions, performance, and results of experimental research are shaped by the available technologies required to perform particular research projects. Physics is premised upon the legitimization and justification of techniques and anything that cannot be disclosed via publicly accepted communicable and repeatable techniques of disclosure are excluded from being included as an object for research. The "objects" of theoretical physics must be both graphically visualised and technologically implemented in order to be intelligible and available objects for experimentation and mathematical modelling. This allows theories to be "tested". How are theories, representation, and experiments linked? Ian Hacking characterised experimentation as a series of interventions and have examined how physicists use representations to plan and order these interventions.[2] Hacking questioned the traditional distinction between experiments that "intervene" and theories that "represent". He argued that theories also intervene as part of the process of making observations in experimental practice and that experimentation is not simply a process of comparing representations with observations because representations are instrumentally required in order to make the interventions necessary for observations in the first instance. I agree with Hacking. Theoretical and experimental practices are linked by making visual representations of "invisible entities" (such as "subatomic particles", "space-time curvatures", "superstrings", and "electromagnetic fields") that can be related to interventions through the mediation of calibrated instruments and apparatus. However, once we take the operational function of the precepts of mechanical realism into account then we can generalise further. Theoretical objects are conceptually linked to technological objects by using a conception of mechanism as the link between natural causes and effects.[3] The internal relationship between theoretical and technological practice is a series of implementations of techniques abstracted as the realization of mechanisms. This permits the progressive development of both theory and technology. Experimental physics is directed towards the achievement of a complete account of the unchanging first principles of causal change in terms of the mechanisms that govern the performances of the experimental apparatus. Such performances are taken to be the natural responses of the apparatus to human interventions and are taken by experimental physicists to disclose natural mechanisms at work. Knowledge of such principles of change have the characteristic form of a theoretical knowledge of how change happens due to the actualization and exercise of natural causes, mechanisms, powers, and structures in accordance with natural law. The successful implementation of a mechanism in technological practice is taken to be the actual disclosure of those otherwise "invisible" theoretical entities as the effects of real entities. These entities are disclosed by technological operations that are interpreted using graphically visualised "natural mechanisms" by which these "invisible entities" supposedly interact via the components of the machinery of the apparatus. The work of novel experimentation is to provide visual representations and intelligible models of "the invisible world" that allow the "objects" and "mechanisms" of theories to become "observable" and available for manipulation by interventional techniques. The only postulated theoretical entities that are acceptable within experimental physics (whether described as particles, forces, waves, or whatever) are those that can be used to interpret machine performance, via a mechanism, and when these are instrumental in the innovation of novel machines then they are taken to be real within the ontology of mechanical realist physics. The epistemological character of scientific knowledge is reduced to "know-how" and the study of natural phenomena is reduced to the search for fundamental mechanisms and their implementation in future scientific work. Thus experimental physics is premised upon the "how does it work?" question and it "tests" any answer to that question by attempting to produce and reproduce the proposed mechanisms in the ongoing work of technological innovation. Hence the truth-status of any theory in physics is perpetually deferred to its future utility in the extension of the ongoing work of experimentation. The operational metaphysics of mechanical realism entails these precepts being situated within ongoing technological practices as an experimental mode of disclosure.

As I shall argue in the next chapter, the "self-evidence" of mechanical realism was established in the sixteenth century. It has become so widely and deeply accepted within the epistemology and ontology of physics, as well as other experimental sciences, that it has ceased to be a part of metaphysics at all and has become an inarticulate and habitual set of beliefs, values, and presuppositions. It has been paradigmatically inherited and embodied in discourse, experience, and practice, via education and the acceptance of scientific authority. This paradigm remains concealed behind a positivistic disciplinary matrix until its metaphysical stepping-stones have been disclosed. Nicholas Maxwell argued that physics is unintelligible, as a human pursuit, unless we examine the metaphysical assumptions that are required for evidence and theories to be comprehensible.[4] For Maxwell, physics is only possible because of speculative metaphysical assumptions regarding the ultimate nature of the Universe and he considered the role of philosophy to be that of revealing these assumptions. Physicists need to make judgements about the ultimate nature of objective reality from a position of complete ignorance in order to begin the inquiry into the ultimate nature of objective reality. Hence Maxwell argued that

> Metaphysics determines methodology. This makes it of paramount importance that a good basic metaphysical conjecture is adopted, one that corresponds to how the universe actually is. A bad metaphysical conjecture, hopelessly at odds with the actual nature of the Universe, will lead to the adoption of an entirely inappropriate set of *methods*, and the result will be failure, possibly, of a peculiarly persistent kind.[5]

He proposed a ten level hierarchy of metaphysical assumptions concerning the comprehensibility of the Universe and demanded that the assumptions implicit in current scientific methodology are made explicit and increasingly simplified. However, he presumed that once we have arrived at indubitable assumptions (that can not be doubted without impeding the growth of knowledge) then we have a good reason to believe that we are nearing the truth. I wish to put aside the obvious philosophical objections to such a presumption. My contention with Maxwell is that he has presumed that physics is self-evidently successful from the onset. This presupposition allowed Maxwell to start with "evidence" as the first level of his ten levels system. Thus he only offered us a system based on principles of simplicity and comprehensibility for explicating the speculative metaphysical assumptions used to interpret "evidence". However, his system does not provide us with an account of how "evidence" is selected and produced *qua* evidence. He did not provide us with any account of the operational metaphysics that underlies experimental inquiry using instruments and apparatus as a means of producing evidence about the Universe. In my terms, Maxwell's system presupposes mechanical realism as its underlying operational metaphysics. This is necessary for the disclosure of a level one "evidence" on the basis of experiments using machines. It is an implicit "level zero" that allows experiments upon machines to be presented as disclosures of natural mechanisms.

Mechanical realism is premised upon and justifies an uncritical acceptance of the neutrality of technology. The mechanical realist believes that technology provides a neutral means by which scientific theories about the connection between the observable and the unobservable can be tested. S/he argues that indirect evidence, through mathematical and experimental practices, can relate the observable effects of unobservable causes with theoretical explanations of such changes in terms of mechanisms. This supposedly provides such theories with explanatory power. Such arguments take the form of "inference to the best explanation arguments" – if a theory explains some "data" better than any other theory, we supposedly have a good reason to think that it is true or approximately true. Explanatory power is taken to be a reason for belief. It is inherent to scientific realism that science progresses by providing increasingly better explanations. The current scientific explanations may not be true but somehow they are truer than the previous explanations. As Popper put it:

> [Scientific realists] not only assume that there is a real world but that also this world is by and large more similar to the way modern theories describe it than to the way superseded theories describe it. On this basis, we can argue that it would be a highly improbable

coincidence if a theory like Einstein's could correctly predict very precise measurements not predicted by its predecessors unless there is 'some truth' in it.[6]

In other words, the predictive success of Einstein's theory supposedly underwrites (at least tentatively) the existence of unobservable entities it postulates. These entities, such as regions of space–time curvature and invariant metrics, are supposedly underwritten by the technological success of using Einstein's theory in the exploratory work of observational astronomy and astrophysics. The scientific realist argues that if a theory based on unobservable entities produces predictions of observable regularities, and expands the boundaries of what can be observed, then what that theory has to say about the unobservable world has a good chance of being true. Given that we have predictively successful scientific explanatory theories we supposedly have a reason to think of them as true, or at least, approximately true. This kind of argument for a scientific realist interpretation of scientific theory acceptance is an argument for *methodological scientific realism*.[7] This scientific realist argument is concerned with what science can and cannot do. It expresses the hope that we can transcend human perspectives and achieve knowledge that will act as a corrective activity upon the social and historical differences in our perceptions of the world into which we are born. Furthermore, through technology, science promises to take us much further than our limited bodies would otherwise allow. We can extend our powers of observation by using instrumentation and send probes to distant planets, for example. It is this very technological power that supposedly provides us with a wider and more inclusive world-view than our inborn nature would provide, and hopefully will provide us with a less relative and more absolute account of the world. This position requires and maintains a distinction between the appearance of the world (according to our sensory experience) and the reality of the world (possibly open to experience through technological powers and scientific exploration). It is this distinction that required the categorization of the world into primary and secondary qualities or characteristics. This position requires the acceptance of the existence of an absolute, complete world, as it is in itself, to which any particular modes of perception could be related, at least in principle. Any such perception would be available to any being of any bodily or contingent constitution provided any such being was able to perform scientific activity and thinking.

This position is a modern manifestation of the ancient dream that everything in the world can be expressed and known as derivative from universal and eternal principles. This is an ideal which, at least as of yet, science has been unable to actually achieve, nor demonstrate that it is achievable in the future. The scientific realist's position is one of a faith that science will be able to provide *episteme* even though any *"epistemoi"* so far provided by scientific activity have been short lived and disputable. It is this faith that reveals the extent that scientific realism, as a philosophical position, is based on a psychological disposition that makes scientific activities and discourses meaningful for methodological scientific realists as being based, if not on knowledge, but on the faith in the possibility of achieving knowledge. However, if we intend to understand the methodology of physics then we need to understand the methodological function of this faith. Do physicists need to be scientific realists? Does a physicist who adopts scientific realism behave differently from one who does not? The methodological scientific realist argues that physicists seek a literal understanding of past and present theories, and the use of concepts underwrites their employment in the construction of new theories. The basis for this belief is that *new* theories point out – and explain – *new* phenomena. This achievement is Bacon's dream of *new* knowledge offering *new* powers. If science is based on scientific realism, and science achieves successes in both prediction and communication, then the methodological scientific realist considers it absurd to claim that science could achieve such successes, being based on scientific realism, without achieving some approximation to the truth. Methodological scientific realism is based upon instrumentality and intelligibility arguments for truth. This presupposes that physics is successful and that success underwrites the realist motivation for being a physicist. Let us also assume that physics does achieve success in making predictions and achieving new powers and that physicists are (by and large) scientific realists. Does it follow from this that the rest of us should be scientific realists? In *A Realist Theory of Science*, Roy Bhaskar argued that it does.[8]

Bhaskar's realist theory of science

Bhaskar argued that material practice is central to the endeavour of experimental physics. He stands alongside many within contemporary science studies, such as Hacking, Gooding, and Pickering, in this respect. Bhaskar also argued that science is a social product and activity. Knowledge is not created in a vacuum out of nothing. It can only be produced by means of production, as revised understandings are achieved via the transformation of existing insights, hypotheses, guesses, and anomalies, etc. Bhaskar claimed that scientists seek to account for some phenomenon of interest in an open system by using a closed system of the experiment to produce a regular pattern of events with the aim of exercising and identifying a (set of) mechanism(s) most directly responsible for those events. Producing this explanation will involve drawing upon existing cognitive material, operating under the control of analogical reasoning and metaphor, to construct a theory of mechanism that, if it were to work in the postulated way, could account for the phenomenon in question. Bhaskar accepted that social activities such as the scientific practices of experimental activity are historically transient and dependent on the powers of human beings as causal agents. He asserted that it is not the task of philosophy to determine what the mechanisms are – that is the task of science – but, rather, "the function of philosophy is to analyse concepts that are 'already given' but 'as confused'".[9] The work of philosophy is supposedly to critically examine the questions "put to reality" and the manner in which this is done. He proposed his transcendental realism to make science intelligible and answer the metaphysical question: what makes science possible? What must the world be like for those scientific practices to be possible? Transcendental realism, unlike Hacking's instrumental realism, Pickering's pragmatic realism, and Gooding's asymptotic realism, is both descriptive *and prescriptive*. It is a philosophy *for* science, rather than merely a philosophy *of* science.

The main thrust of his argument was against positivism. He argued that positivism is unable to sustain either the necessity or the universality of natural laws because positivism, whilst affirming the existence of things, events, and/or states of affairs, denies the possibility of any knowledge of underlying causes, powers, or structures. Positivism restricts knowledge to statements such as "when A happens then B happens" but does not claim "A causes B". He argued that positivism is guilty of "an epistemic fallacy" by reducing the question of "what is" to that of "what can be known" and cannot provide an intelligible theory of experimental science. Bhaskar declared his intention "to provide a comprehensive alternative to the positivism that has usurped the title of science" and proposed his transcendental realist theory of experimental science as a third position to stand against both positivism and idealism.[10] He distinguished his transcendental realism from positivism by arguing that a pattern of events produced in experiments signifies the existence of an invariant generative or causal mechanism rather than merely signifying regularity. He distinguished his position from idealism by allowing the possibility that generative or causal mechanisms referred to in explanations as something that may be real rather than always imaginary. However, Bhaskar went further than just providing a realist interpretation of experimental physics. He claimed that a realist interpretation of science is *necessary*, if experimental activity is to be intelligible. He argued that his transcendental realism "is the only position that can do justice to science" because "without such an interpretation it is impossible to sustain the rationality of any scientific growth or change".[11] He maintained that there is an ontological distinction between scientific laws and patterns of events because the core of theory has a conception or a picture of a natural mechanism or structure at work, and asserted that natural laws operate independently of the conditions for their identification in terms of patterns of events. Given that in the world outside the laboratory walls is a complex place in which regularly repeated constant conjunctions (or regular patterns) of events (such as "when A happens then B happens"), required by the positivistic sciences, are uncommon, then a positivistic science would only operate within the circumscribed confines of the closed system. It would not be able to justify the production of any conjunctions of events that it was able to produce because, as empirical regularities, they would not be transferable from the context of production. These empirical regularities could not be presented as the consequences of natural laws. Thus, for Bhaskar, empiricism cannot make the practices of experimental sciences intelligible as a human pursuit. According to Bhaskar, experimental physicists seek to isolate mechanisms and do not stop at noting the existence of constant conjunctions. The presupposition of the existence of a law must be prior to any attempt to identify it. For example, for Faraday and Maxwell's efforts to be intelligible, the laws of electromagnetism must be existent prior to and independently of Faraday's experiments on electromagnetic phenomena and Maxwell's mathematical formulation of those laws. The purpose of experimentation is to actualise the mechanisms that are governed by those laws and any intelligible account of experimentation must presuppose that natural laws are prior to and transcend the experimental activity that actualises them. Thus, for Bhaskar, the assumption of the efficacy of a law must precede the attempts to actualise and stabilise a pattern of events if those attempts are to be intelligible. If the laws of electromagnetism, for example, are real independently of whether they are known to exist then we can presume that they exist prior to their actualization in the movements of the magnetic needle when a magnetic compass is moved next to an electric wire. Furthermore, we can also presume that the actualization of those laws occurs, under those conditions, whether or not we empirically observe the compass needle move. Hence, his argument for the realist structure of experimental activity was based on the following premises:

(i) Causal laws are ontologically distinct from patterns of events;
(ii) Causal laws are those aspects of reality which underpin, generate, or facilitate the actual phenomenon that we may (or may not) experience;
(iii) The intelligibility of experimentation presupposes that reality is constituted not only by experiences and the course of actual events, but also by structures, powers, mechanisms, and tendencies;
(iv) Knowledge cannot be equated with direct experience;
(v) An adequate account of science requires the presumption that reality revealed by science exists independently of the human efforts to reveal that reality.

His conclusion was that only a realist analysis could sustain the intelligibility of using artificially closed systems in experimental science to learn about the natural processes in the open systems of the world. When an experiment has been set up so that only one mechanism, or a single set of mechanisms, operates then we have a closed system. Even though no system is ever perfectly closed, experiments can approximate a closed system sufficiently enough to satisfy the purposes of experimental science. The reality of the mechanism(s) disclosed by experimental activity is subsequently subjected to empirical scrutiny and the empirical adequacy of the hypothesis is compared to that of competing explanations. The concept of a mechanism provides a metaphysical conceptual link between change in the real world and changes in machine performances in the laboratory. A mechanism is an index for the repeatable processes and machine performances that provides a central discursive function within causal accounts used to explain those processes and performances. On Bhaskar's account, mechanisms can only be identified in closed systems and are restricted to the artificial contexts of experimental sciences, but can be used to explain the events in the "open systems" of the real world. Hence, Bhaskar described an experiment as "an attempt to trigger or unleash a single kind of mechanism or process in relative isolation, free from the interfering flux of the open world, so as to observe its detailed workings or record its characteristic mode of effect and/or test some hypothesis about them."[12]

Bhaskar argued an adequate theory of experimentation must allow for three kinds of "ontological depth": *intransitivity*, *transfactuality*, and *stratification*:

Intransitivity allows the possibility that the mechanisms identified by experiment operate prior to and independently of their discovery, and also that they are not changed by the processes involved in their discovery. Whilst the transitive dimensions of experimental activity are the contingent practices of the experimenters and the current theories that they are working with, the intransitive dimensions are the mechanisms and laws that the experimenters are trying to discover.

Transfactuality allows the possibility that "the laws of Nature" exist and operate independently from the closure of the systems in which they also can be existent and operational in the artificial production of empirical regularities. The constant conjunctions of events in experiments are produced in accordance with natural laws that operate independently of the experiments that disclose them. Theoretical explanations explain laws in terms of the structures which account for constant conjunctions in closed systems, while they are applied transfactually in the practical explanation of the phenomena observed in open systems. Bhaskar proposed that his analysis provided a condition of the intelligibility of experimentation because the laws that science identifies under experimental conditions continue to hold as transfactuals (not empirical regularities) extra-experimentally. This is the rationale for practical, explanatory, diagnostic, and exploratory scientific work. The purpose of experimenting is to identify a universal law (within its range), which, in virtue of the need to perform a disclosing experiment, is not actually or empirically present, and allows experimental science to operate within a context of discovery. Once the ubiquity of open systems and the necessity for experimentation are appreciated, then laws must be analysed as transfactual, as universal (within their range), but neither actual nor empirical.

Stratification allows the possibility that Nature imposes a certain dynamic logic to scientific discovery, in which progressively deeper knowledge of natural mechanisms is achieved, as "the strata of reality" are uncovered *a posteriori*. Bhaskar claimed that this conception of stratification allowed him to isolate a general dynamic of scientific discovery and the development involving the identification of different strata of explanations given in terms of natural necessity, tendencies, and mechanisms. For example, the thermodynamic mechanisms invoked to explain the heating of water can be explained in terms of the bonding and elastic properties of molecules of water, which, in turn, can be explained in terms of the electronic structure of water molecules, and, in turn, the principles of quantum mechanics, and so on. The historical development of the scientific understanding of why water boils when heated (rather than freezing) in terms of "deeper strata" of explanation follows "the strata of reality" because it is necessary to explore higher-order strata before one can investigate deeper ones. Bhaskar argued that it follows from the stratification of reality that any adequate science must provide stratified explanations in terms of a multiplicity of levels of causal mechanisms. Even though one kind of mechanism may be explained or grounded in terms of another, it cannot be necessarily reduced to or explained away in terms of it. Such grounding is consistent with its emergence, so that the course of both Nature and scientific exploration are different than it would have been if the more basic stratum alone operated. The higher-order structure is real and worthy of scientific investigation in its own right. Any explanation that is (tentatively) accepted must be also explained and this further explanation must in turn be explained on the basis of a stratification of reality. The real multiplicity and stratification of natural mechanisms grounds a real plurality of sciences that study them.

At the base of these three kinds of "ontological depth" was a causal criterion for attributing reality. The theoretical entities and processes proposed as plausible explanations of observed phenomena were to be established as real through the construction either of sense-extending equipment or of instruments capable of detecting the effects of phenomena. He presumed that accurate models in terms of functionality necessarily imply that they are accurate models of the phenomenon under investigation and, consequently, these models disclose the underlying mechanisms of the phenomena under investigation. These three kinds of "ontological depth" also provided two criteria for an adequate realist philosophy of science. Bhaskar termed these two criteria as *vertical* and *horizontal* realism. *Vertical* realism assumes that science is a progressive, continuous, and reiterative process of movement from manifest phenomena, through experimentation and creative modelling, to the identification of generative causes, which then become the new phenomena to be explained. *Horizontal* realism assumes the universality of the workings of generative laws or mechanisms (within their range) and is a statement of the independence of the mechanism discovered during experiments. The horizontal aspect supposedly explained why mechanisms may be possessed unobserved and the vertical aspect supposedly explained why they could also be discovered in an ongoing irreducibly open-ended process of scientific development. This two-fold realism is committed to the belief that the powers or tendencies of underlying generative mechanisms disclosed through experimentation provides us with insights into the nature of things outside the context of the experiment. Without this belief, for Bhaskar, experimental science is unintelligible and an essential feature of experimental science is that it uses the closed system to find the otherwise hidden mechanisms that operate in open systems as a means of producing causal explanations about the phenomena in open systems. Therefore it cannot be accounted for in terms of positivism because, as he put it, "scientifically significant generality does not lie on the face of the world, but in the hidden essences of things."[13] The process of science is to move from the identification of "protolaws" to the identification of "laws of Nature", where a "protolaw" is a potentially non-random patterns and results of Nature (including in the laboratory) that are epistemologically significant. The crucial scientific transition is the identification of a generative mechanism or structure that explains a "protolaw" and would ground a law of Nature i.e. a transfactual and efficacious tendency, understood as universal (within its range) but non-empirical, necessary, and discovered *a posteriori*.

The inadequacy of empirical adequacy

Many historians and philosophers of science, such as Kuhn, Feyerabend, Hacking, and Gooding, have shown that the boundary between the observational and the theoretical is ambiguous – it may change with new instrumentations, concepts, and representational techniques. New experiences are made possible through the innovation of instruments such as the electron microscope and the microwave telescope. The innovation of new instruments has led to the growth of "the observable realm" and, as even Popper accepted, observations are made in the light of theories.[14] Bhaskar and Harré not only accepted that the act of making an observation is guided by theory, but also argued that it was an essential aspect of the intelligibility of experimentation that it was. However, they did not explain how theories and observations were to be connected in the ongoing work of experimentation. Gooding has provided a clear and detailed account of how Faraday made and used visualizations and interventions within the ongoing work of constructing and communicating his experiments.[15] The positivistic restriction of the ontology knowable through experimental science to that of observable entities is extremely problematical because of the complex relations between theory and observation when instruments mediate them. It is not simply a matter of passively seeing what is there because one must also interpret what one sees. It was for this reason that Kuhn and Feyerabend argued that observations using instruments are "theory laden", and, as Hacking and Gooding argued, observations are made in experiments through a series of interventions using pre-theoretical pictures and many observations are not even formally "theory laden" but are made using tentative visual representations. Gooding argued that experimenters make a tentative visualization, or construal, of a phenomenon before making a theoretical interpretation of it and use these construals to show others how to make the same observation and to guide further interventions. The indirectness of perceptual experience when using such instruments is a serious problem for a positivistic conception of science because the processes involved in using novel instrumentation to make novel observations are not based on empirical and logical propositions.

This problem is made worse for the positivist if the initial innovation and development of a new instrument, such as a solar neutrino detector or electron microscope, is based upon a theory that utilises unobservable mechanisms, such as neutrino oscillations or electronic quantum tunnelling, and unobservable entities, such as neutrinos and electrons. The shift in the boundary between theoretically possible observations and "the observable realm" is navigated using instruments that are imagined, designed, built, and operated through the implementation of theories, interpretations, and construals based upon postulated entities and mechanisms that cannot be directly observed. The construction and operation of instruments such as electron microscopes, x-ray scanners, neutrino detectors, Geiger counters, microwave telescopes, etc., is at best highly problematic for the positivist and at worst is meaningless and unscientific. We do not have a direct perceptual acquaintance with "objects" such as electrons, genes, and electromagnetic waves, etc., at any time or in any context. We infer their existence on the basis of the use of theoretical interpretations, representations, and construals of the outputs and performance of machines. Pointing at a tree and saying "tree" is not the same type of reference-act as pointing at a test-tube full of gloopy liquid and saying "genes", pointing at a photograph of slightly curved lines and saying "electron", or pointing at the change in the position of a magnetic needle next to a electric wire and saying "electromagnetic wave".

Inferences are not the same type of reference-act as perceptions. The act of looking at the output of an electron microscope and seeing the internal structure of bacteria, or the act of looking at the output of a radio telescope and seeing a distant quasar, requires specialised interpretation and training to be able to perceive "what is there" in terms of a theoretical understanding of how the detector works. The ontological status of "causal mechanisms" and their role in the construction of experience is crucial for an understanding of the relations between theory and experience in experimental physics. Do we need to be realists about "causal mechanisms" when the construction and use of instruments in experimental physics is initiated and understood in terms of "causal mechanisms" and experimental physics is shown to technologically progress and achieve predictive accuracy by using those instruments? As Hacking insisted, along with the methodological scientific realists, the growth of the observable realm through instruments is an important feature of science. If the same entities can be observed with independent instruments then there is a good reason to believe that they exist independently of those instruments. This belief in the transfactuality of the existence of theoretical entities, such as electrons or electromagnetic fields, is central to the intelligibility of experimental physics as a pursuit and emphasises the importance of cross checking. Thus the instrumentality of any theory in successfully making different experiments and observations supposedly provides a good reason to believe that the theoretical entities exist. This point was also central to Harré and Bhaskar's arguments for scientific realism. The prediction of possible novel phenomena is an important aspect of theoretical work. Physicists use theories to *discover and explain* the elements of the world that exist *independently* of human experience and, for the methodological realist, the rationality of this intention is epistemologically justified by their successes in achieving it. If novel observations and powers are made possible because of a new device then it would seem that there are good reasons to believe that the theoretical causal mechanisms utilised in the conception and design of that device must exist.

However, the situation is not quite that simple. Scientific theories are generalizations, idealizations, and abstractions, which focus on particular properties of phenomena under investigation and cases of partial regularity, that are modified in accordance with the particularities of use in context. Hence, as Nancy Cartwright argued, that literal representation is not a criterion in theoretical modelling.[16] Assumptions and approximations are accepted which, although not exactly true, are not exactly false either. Physicists simplify complex natural phenomena in order to provide the simplicity appropriate for manipulation and modelling within the context of the experiment, and theories are corrected and modified in order to fit the facts. Thus both theories and the objects to which they apply are constructed and then matched, in a piecemeal fashion, to the real situations. This process sometimes provides predictive accuracy but rarely do theories and models match all the facts at once. Hence, Cartwright argued that fundamental laws do not, in fact, describe reality but only describe the appearance of reality that "is far tidier and more readily regimented than reality itself."[17] Physicists can only achieve abstract descriptions of "the appearance of reality" and these are insufficient to provide good reasons for a belief in the theoretical entities used to produce those descriptions. It is possible that novel machines could be built on the basis of an instrumental and piecemeal use of a collection of theories, construals, representations, and models as heuristic guides. The attempts to produce a neat, consistent, unified, and abstract theoretical understanding of novel machines often retrospectively follow the innovation and development of that machine (as in the case of the steam engine or the electromagnetic motor). Such theories could always be replaced in the future and, consequently, even though the empiricist will argue that theories must accurately describe the behaviour of phenomena, this does not give us compelling reasons to believe that the theoretical entities postulated in empirically adequate theories really exist.

In her later work, Cartwright argued that natural laws and the scientific theories that describe them are not universally applicable, but are true only *ceteris paribus* and the content of our best scientific theories describes the capacities things have as a result of possessing certain features or properties.[18] She argued that even the best descriptions of natural laws are highly limited in scope and apply only to very specific arrangements of things that the regularities described by these laws. She termed these very specific arrangements as *nomological machines*. She defined a nomological machine to be an arrangement of objects and properties that have stable capacities or powers that generate empirical regularities.[19] Such a machine may be very simple, such as a rigid rod placed on a fulcrum that serves as a lever, or it may be very complicated, such as the Stanford Gravity Probe. The important feature of such machines, according to Cartwright, is that they possess capacities that generate regular behaviour when the machines are set running in the right conditions. The right conditions include the fact that the machine is shielded from unwanted outside causal influences that would interfere with its operation. Regularities are hard to find because nomological machines operating in shielded conditions rarely occur naturally and it requires experimental intervention and control to shield a nomological machine in the right way for it to generate the appropriate regularities. Cartwright argued that capacities are more ontologically fundamental than empirical regularities because the existence of such regularities depends upon the existence of the capacities of a nomological machine. Laws only hold so long as such arrangements are in evidence and effectively shielded from interfering factors. Scientists construct models of for the various capacities of the arrangement in specific circumstances, but in other circumstances different combinations of capacities may be evidence and a different law may be applicable. As a consequence of this, Cartwright argued that laws do not account for the behaviours of things in the world because specific laws hold only insofar as things with stable capacities are thrown together in appropriate circumstances. In her view, we need to examine the nature of capacities if we are to understand how things behave.

Cartwright's criticisms were primarily directed towards the assumption that laws are universally applicable to all domains. Contrary to this claim, Cartwright argued that unless we have a model for a law, which provides correct predictions within experimental tests, then we do not have an empirical basis for the belief that the law applies. For example, on Cartwright's account, whereas we have good reasons to believe that Newton's Second Law of Motion may well be applicable to the motion of a gyroscope in space, because we have an accurate model for this, we have no such model for accurately calculating net forces on a banknote blowing in the wind, and, consequently, the belief that Newton's Second Law applies in merely the prejudice of a fundamentalist. In the absence of an empirically adequate model for the phenomenon then we have no empirically based reason to think that Newton's Second Law applies to the movements of a banknote in the wind. Her point is that if the situation is too complex to model and experimentally reproduce as a normative machine then we cannot know whether this abstract law is relevant or there is a different law describing different capacities in that situation. The law and model are both empirically limited to the case of the nomological machine in question and, hence, we do not have any *a priori* reason to claim that the law applies beyond this machine (or set of machines). Cartwright demanded that, if we aim to be empirical, that we should reject the validity of the fundamentalists' extension of the remit of the law to particular complicated situations for which we lack any empirically adequate model, and we should accept the patchwork and discontinuous nature of the Universe open to scientific investigation. However, even though Cartwright raised some interesting critical points about the ontology disclosed by experimental physics, her argument neglected to address the concept of "mechanism" and the use of such a concept in extending and combining the models used in the laboratory to the wider world. Such a concept is essential to connect the nomological machine with the complex phenomena of the natural world. Cartwright neglected to explain the rationale of using nomological machines to explore natural phenomena because she did not explain how experimenters connect their efforts in the artificial circumstances of the laboratory to the phenomena of the natural world. Her interpretation of science is unable to explain the scientists' rationale for using experimental normative machines to explore complex phenomena that they claim to pre-exist their efforts to explore them. As Bhaskar argued, a concept of "mechanism" is essential to connect the closed system of the empirical regularity producing experiment with the open system of the natural world because, without such a concept, the classification of experimental physics as a natural science is completely unintelligible and arbitrary. Without a concept of "mechanism", scientists would be unable to explain how the empirical regularities were produced during an experiment, either in terms of tendencies or capacities, nor would they be able to extend those explanations to the natural world. The intelligibility of any model is independent from its truth-status; the ability of physicists to bring transformative powers into the world does not logically support the truth-status of their causal accounts. Unfortunately for Cartwright's whole argument, it does not follow from the fact that predictable models are empirically limited to machine performances of particular machine-kinds, or nomological machines, that the Universe is shown to actually be a patchwork of such machines and that physicists should not treat natural laws as universal. Instead, Cartwright's analysis of experimental science reveals the extent that empiricism is an inadequate philosophy of science and that the natural science of experimental physics is not actually an empirical science (in the positivistic sense).

However, it does not immediately follow from pragmatic successes using particular theoretical understanding and prediction of the empirical results of an experiment, given in terms of particular mechanisms, that any such mechanisms exists independently from the particular experiments they are used to explain and predict. There is always the possibility that at the empirical level two different theories can agree but utilise different theoretical entities and mechanisms. This plurality suggests that there are not any necessary relations between any explanatory intelligibility, predictive success, and objective truth. It is possible that through trial and error, using theories only as a heuristic resource, we could eventually build a device that provides new powers. It is possible that we could build that device without understanding how it works, or we could build that device using one theory and subsequently completely re-evaluate our theoretical understanding of that device according to another theory. As Bas van Fraassen argued, many theoretical entities, utilised in past theories, have become merely of historical interest whereas the empirical knowledge those theories have produced have often remained.[20] Newtonian absolute space and time have been replaced with the Einstein's relative space-time as a theoretical entity due to the empirical inadequacy of Newton's theories of motion and gravitation. Yet much of the empirical adequacy, which Newton's theories produced, remains and is also obtainable using the more empirically adequate Einstein's theory. If past theories have been shown to be empirically inadequate then this provides a good reason to suppose that current theories will be shown to be empirically inadequate in the future, and the literal truth of any theory does not follow from the current empirical adequacy of that theory. It is a logical possibility that Einstein's theory will be in turn replaced with a more empirically adequate theory that will use different theoretical entities. Even if we accept that scientific theories are progressing in their empirical accuracy, predictive power, explanatory success, and productivity of new phenomena, this still does not provide us with any certainty that the theories that are being currently used by working scientists will not be replaced by subsequent theories. In fact, this acceptance of the progress of scientific knowledge prohibits any such certainty. As a consequence of this, van Fraassen argued that it logically follows from such a prohibition that not only is there reasonable doubt in the theory-independence of unobservable entities, but also that there is no pre-requisite for any *isomorphism* between the structures of a theory and the structures of the object of that theory, for it to be successful at an empirical level.

The epistemological standard for any adequate scientific theory is whether it matches empirical appearances, and this evaluation is made independently of its ontological commitments. Hence, van Fraassen rejected scientific realism and its ontological commitments in favour of that given as a matter of direct observational experience. This involves adopting the traditional empiricist scepticism regarding the existence of any objective causes or entities that are not apparent as actually observable phenomena. Scientific theories arise through competitive social processes in which only the successful theories emerge and we should not be surprised at the predictive and descriptive successes of current scientific theories. These theories are latched onto observed regularities in Nature.[21] Furthermore, he claimed that the demand for causal explanations should not play an epistemological role in "the scientific enterprise".[22] He adopted the instrumentalist line that the best that physicists could legitimately claim to achieve are empirically adequate descriptions, predictive success, and manipulative control. The best that physicists can achieve, if the evidence is in favour of that theory, is that the evidence suggests that things behave as if unobservable theoretical entities existed. This does not support any claim that those entities exist and physicists should rest content with saving phenomenal appearances and reject all explanatory causal-accounts, the notion of objective Nature, and claims concerning the existence of unobservable entities. Allegedly, objective and mind-independent reality is, by definition, beyond the capacities of human understanding and, therefore, the task of producing a complete and correct theory of objective reality is an impossible one.[23] The physicist should be content to adequately describe phenomenal appearances by producing abstract formalisms and sets of equations that successfully predict observational results. Explanatory causal-accounts amount to nothing more than fictions.

Empirical approaches such as van Fraassen's maintain that there is a clear distinction between description and explanation. Explanatory theories aim to represent "the causes" of experience, whereas empirical descriptions aim to represent, clarify, and structure experience. The "world of our experience" is irreducible to causal explanations and consequently there is not any necessary connection between any causal account and our experience of the world.[24] Hence, for the empiricist, the indubitable facts of experience have an epistemological primacy over causal explanations. For example, our experience of the world contains phenomena such as blue skies on clear summertime days. How would a physicist explain such a phenomenon? How would a physicist explain the experience of seeing the colour blue to someone who had been blind since birth? Let us assume that the blind person is conversant in the language of modern physics. The physicist could explain the eye in terms of an optical device. S/he could explain how electromagnetic waves of a particular wavelength radiate from the sun, are refracted and scattered by particles in the atmosphere, are focused onto the retina by the lens of the eye, stimulate the rods and the cones of the retina, and are transformed into electromagnetic pulses in the optic nerves. S/he could then explain how these electromagnetic pulses travel through the optic nerves, travel through a network of nerves leading to the brain, generate electrochemical process in the brain's network of neurones, and are finally processed by the brain as the colour blue. Let us assume that whatever theory or model the physicist utters is empirically adequate to the extent that its derivative resultant would be that the sky on a clear day (quantified in terms of humidity, pressure, and temperature) would have the colour blue (quantified in terms of wavelength). Let us also assume that the blind person perfectly understands the physicist's explanation of how the eye and brain processes differences in wavelength. But does the blind person know, on the basis of this explanation, what an experience of the colour blue is? There is one essential characteristic of the colour blue that is missing from the physicist's description of blue. That essential characteristic is the quality of clear daytime sky, *the blueness*, which is immediately experienced by all people who are able to see it. The blind person will, from the physicist's explanation, have no idea whatsoever of what the experience of seeing the colour blue is, nor that there is even the possibility of experiencing blueness. The blind person would have no more idea of the experience of looking at the sky and experiencing blueness, as a result of the physicist's explanation, than s/he did at the onset. The physicist could talk of atoms, electrons, photons, matter, ions, radiation, wavelengths, refraction, the spectrum, prisms, electromagnetism, oscillations, coupling-constants, resonance, or whatever, but would be unable to introduce the quality of blueness into her/his explanation. Of course, the physicist could attempt to explain away the blueness of the sky on a clear day as a subjective illusion, but any arguments against the fact of the blueness of the clear daytime sky require assumptions and premises that would be more suspect than the indubitable experience of blueness. Blueness remains *surplus* to the physicist's explanations of the colour blue. It lies outside the language of physics and remains undeniably *residua* to any attempt to explain it away. The same is also true of all the qualia that are characteristic of the lived-world of human experience.

The empiricists' point is not that human experience is an existential condition for the existence of the world, but, rather, that what we can legitimately say about the world is limited by what we know about that world based on experience. If descriptions and explanations of experience are independent from one another then we cannot claim to have knowledge of causes based upon knowledge of the facts of experience. Hence the realist ontology of science is not supported by the empiricist epistemology of science. However, studies of the behaviour of familiar objects, at most, constitute starting points for experimental physics. The majority of experimental work in experimental physics involves the investigation of phenomena of which we have no direct experience whatsoever. The studies of the majority of the phenomena investigated by the studies of mechanics, thermodynamics, electromagnetism, radioactivity, solid-state physics, and quantum physics involves the interpretation of the performances of machines and instruments. We only have experience of the numerical and analogue readings on calibrated meters, oscilloscopes, graph plotters, gauges, computer displays, and other instruments. However, these experiences are not direct sensory experiences. We need the mediation of technical education before we can make sense of our direct sensory experiences of instruments. Each instrument is calibrated using technical entities (such as potential difference, time-signals, inductance, capacitance, thermal capacity, electrical resistance, phase, frequency, mass, magnetic field strength, force, harmonics, electric charge, power, etc.,) quantified in terms of arbitrary SI units (such as kilograms, metres, candelas, seconds, amperes, moles, radians, kelvins, newtons, coulombs, tesla, watts, etc.) These technical entities and calibration units are meaningless outside of the theoretical and technological frameworks in which they occur. By claiming that measurement using such instruments provides direct experience, the empiricist merely has passively accepted the stable results of an historical struggle of technological and theoretical efforts, and the current state of instrumentation, education, and interpretation as givens. The empiricist has neglected the role that theories and causal-accounts, involving theoretical and technical entities, have had in the construction of those instruments and their interpretation. For example, the observation of the empirical regularity that hot objects in a cooler environment cool down, and cold objects in a warmer environment warm up, (formally abstracted and universalised as the Zeroeth Law of Thermodynamics) constitutes an almost trivial starting point for the study of thermodynamics. The physicist wants to know why this empirical regularity occurs. There is a big imaginative step from this "empirical regularity" and an interpretation of it in terms of flowing heat. An even greater imaginative leap is required to quantify heat and construct an apparatus to measure the conservation of that heat flow. On van Fraassen's account, it is hard to see how thermodynamics could have progressed from the Zeroeth to the First Law, let alone how the Second Law, with its esoteric definitions of work and entropy, could be part of "the scientific enterprise". If we were to describe the objects of investigation and theories as fictions, as does van Fraassen, that are instrumentally used to interpret the empirical facts of measurement, we would still require theoretical training and interpretation in order to ascertain what the empirical facts of measurement were (in all but the most trivial of experiments.)

Entities such as potential differences are no more, nor less, fictional than entities such as electromagnetic fields, given the fact that we have no direct experience of either and both are only meaningful within theoretical interpretations. The former are only considered to be more concrete than the latter on the basis of a passive and uncritical acceptance of past theoretical interpretations used to build the apparatus and an arbitrary scepticism about theoretical interpretations used to interpret the performance of the apparatus. If we were to consistently adopt van Fraassen's empiricism then experimental physics would be a process of instrumentally using fictional entities to investigate other fictional entities. If the measurable variables used to produce descriptions in physics were not made in reference to "objective properties" then it would be pointless to map out the empirical variation of the fictional entities used to make measurements intelligible against the fictional entities used to make observations intelligible. The empirical adequacy of any laws that would be produced through such a process would be the pointless interrelation of different kinds of fictional entities. Modern experimental physics could not be grounded on "constructive empiricism" and physicists would not rest content with it. As J.K. Feibleman pointed out, modern experimental physics is not empirical in the philosophical sense; a physicist does not rely upon direct experience but investigates the disclosure of interpreted instrumentation to direct experience.[25] An intelligible account of physics must provide an account of how this disclosure is possible and how it is achieved. Empiricists cannot provide us with an account of how observations of using instruments are obtained because they cannot consider the technical reasoning (made in terms of theoretical entities and causal-accounts) used in the interpretation of measuring instruments and experimental apparatus, as scientific. Thus they either cannot provide a meaningful account of observation within experimental physics in terms of scientific facts, or it is only able to provide accounts of observation by arbitrarily, passively, and uncritically accepting particular technical interpretations of particular measuring devices as given.

Causal-accounts, in terms of theoretical entities and mechanisms, are central to the interpretation of the possibilities of experimental apparatus and measuring instruments during their design, construction, operation, and interpretation. Any causal-account is, at least in part, an explanation. Modern experimental physics could not proceed without them and any physicist who considers him or her self to be an empiricist (in van Fraassen's sense) has disingenuously presupposed realist interpretations in the design, construction, and operation of their experiments. A physicist uses causal accounts to construct an apparatus that relates variables and s/he cannot operate experimentally by restricting the technological discourse to the level of "when substance A is placed near the Geiger counter the current output increases by B" when designing the experiment in the first place. S/he utilises a discourse at the level of "if radioactive material emits energy sufficient to ionise a gas then this will be manifested as a current across a sealed tube with a voltage across it. Therefore we can build a device to measure levels of radiation." The capacity of physicists to construct experiments is dependent upon such causal accounts. Constraining "the empirical" to the confines of the laboratory, as orthodox empirical accounts of science have done, as a means to verify or falsify theories by testing their predictions and deductions, still does not provide us with an account of how those experiments are constructed. Empiricism tends to only account for experimentation in terms of a ready-made procedure of measurement, without any justification of that procedure, because it has neglected to attend to the role of technology in experimentation. Not only do observations actively occur in the light of theories, within theoretical frameworks or paradigms, using concepts, visualizations and construals, but also theoretical accounts are implicit in the design, construction, operation, and interpretation of the instruments and apparatus used to make those observations. The theories and practices of modern physics presuppose representations of the "causes" of experience in their descriptions of the phenomena of experience because they require interpretations of the experimental apparatus and measuring instruments in terms of causal explanations. Once we recognise that the descriptions produced by modern physics make reference to and utilise causal explanations, then we must also recognise that experimental physicists, whether or not they realise it, need to presuppose scientific realism at some level. Van Fraassen's *a priori* rejection of the role that causal accounts play in "the scientific enterprise" of modern experimental physics prevented him from producing an intelligible account of physics. By asserting that sensory observation founds all genuine knowledge, by presuming that only those statements that are derived from experience are legitimate or scientific, the positivist has neglected to attend to the ways that experiences and the means of their production are constructed. What is the phenomenological character and ontological status of the phenomena explored by making experimental machines to disclose them? In order to question their ontological status, we need to examine the complex relationships that occur, through the interface of measuring instruments and experimental apparatus, between experience and technical causal accounts. How are these technical accounts used to make experience?

A robust and clear distinction between explanation and description is not apparent in actual experimental work. Empiricism cannot provide an account of the feedback processes between scientific theories, technical accounts, the construction of experiments, and experience. It is unable to do this because it is premised upon completed knowledge, reasoning, techniques, experiments, and experiences. It is unable to cope with novel (or revolutionary) experiments and cannot explain how and why experiments were constructed in particular ways. Classical empiricism fails because the technological environment of the laboratory is unintelligible to primitive experience. Scientific instrumentation is a meaningless maelstrom of flashing lights, moving pointers, digital displays, graphs, and readouts to the uninitiated. Technological discourse is required to construct a meaningful experience of experimental apparatus and instruments. There is nothing primitive about voltages, time signals, amplitudes, temperatures, and pressures. Empiricism cannot explain how theoretical entities "transcend" neither the particular contexts of their production or use. It also cannot explain how predictive success is possible. This is also the central problem with van Fraassen's constructive empiricism. It is unable to explain how theories based upon fictional entities can provide predictive success and facilitate manipulative control at all. If electrons and photons are merely fictional entities then how can instrumentalism explain the incredible predictive accuracy of Quantum Electrodynamics? Furthermore, if theoretical entities do not have any transfactuality then the validity of any theoretical entity is restricted to its utility within the context of its use. When physicists experiment on electrons in electric circuits they are not operating with the same kind of electrons "annihilated" in the LEP ring at CERN, those "deflected" in Thompson's cathode ray tube experiment, those "suspended" in Millikan's oil drop experiment, or those "sprayed" on Morpurgo's mobidium spheres to detect free quarks. Van Fraassen is unable to explain how these physicists are able to use the same theories, say QED, Special Relativity, or Maxwell's equations, in these distinct contexts and by limiting knowledge to the empirical, it is unable to legitimately transcend and unify the particularities of experiences. Furthermore, van Fraassen cannot consistently provide an account of how physicists fail to successfully perform experiments according to theoretical expectations – except by asserting that it is only a matter of time before all theories demonstrate their empirical inadequacy. Given that Morpurgo's famous experiment to attempt to find free-quarks was a project that utilised fictional entities, electrons and electromagnetic fields, to attempt to find other fictional entities, namely free-quarks, then there does not seem to be any source of resistance to its success. If the proper use and identity of theoretical entities is completely determined in context then there is nothing to prevent a physicist, in that context, from achieving success. One need only instrumentally reconstruct the meaning of the performance of the apparatus according to one's intentions within that context. Other physicists, in other contexts could not object to this because one could merely argue that they were using the terms "electron", "electromagnetic field", and "free-quark" differently within different contexts, and that anyone who failed to repeat the observation merely had failed to re-construct the context of use. Yet Morpurgo did not observe any indication of the existence of free-quarks during his twenty-five year search for them. Any intelligible account of experimental physics must provide an account of how theoretical entities achieve transfactuality and also how experiments fail to perform according to theoretical expectations.

In general, if we accept that the use of causal accounts is an essential part of designing and building apparatus in experimental physics, then empiricism, by rejecting the possibility of the knowledge of causes, cannot provide an intelligible account of how and why novel experiments are constructed and performed. Experimentation is based on planned action, selections, and decisions within the context of the innovation and ordering of material and theoretical practices. It is a social and interactive process in which experiences are made through deliberated material practices rather than passively received. Positivism is unable to deal with the processes of constructing novel experiences of novel phenomena because it is unable to account for the ways that novel material, communication, and visualization practices are constructed when dealing with novel phenomena, nor can it provide an account about how novel experiments can be constructed in the absence of an established theory. It fails to recognise that exploration in experimental work requires speculative and constructive processes of visualization and communication that cannot be reduced to logical or empirical statements.[26] By reducing physics to the systematic logical ordering of our sensory experiences, positivism restricts "facts" to be statements of our immediate experiences made in terms of already given language. Thus it is impossible for us to add any fact that cannot be expressed and logically analysed in terms of already given language. It is impossible for positivists to explain how novel experiences could be constructed in terms of novel expression because the restriction that they place upon legitimacy and intelligibility would not allow the processes involved in the construction of novel expression to begin. Thus positivism cannot account for scientific change nor can it account for how experimental physics began in the first place, and, as a consequence, novel experimental research programs could not proceed according to a rigid positivistic conception. It is only able to provide a conception of experimental physics during its Kuhnian "normal science" phase. Positivism, by uncritically making its appeals to "the authority of scientific experience", rests upon a reactionary and conservative appeal to already completed scientific results.[27] Novel phenomena require new concepts in order to have new experiences, and the application of new concepts implies speculative metaphysical assumptions regarding the nature of the phenomena.

Positivists presume that physics has a logical and empirical character that is divorced from metaphysics. Empiricists, positivists, and instrumentalists reject metaphysics as either nonsense or unverifiable because, by positivistic definition, metaphysics is taken to lack any empirical content. How can positivism reject metaphysics and affirm the physicists' claim to discover or use unobservable entities, laws, or mechanisms? It cannot. A positivistic interpretation of physics is a rejection of the validity of such claims and, as a consequence, must reject the epistemological validity of experimental physics. Furthermore, as Popper noted, various speculative metaphysical beliefs, such as Kepler's harmonics of the geometry of the planetary orbits about the Sun, have lead to significant advances in theory.[28] History provides many examples, such as seventeenth century atomism, that were untestable using the technologies available at the time. As Popper put it:

"... [There] is a least one philosophical problem in which all thinking men are interested. It is the problem of cosmology: *the problem of understanding the world – including ourselves, and our knowledge, as part of the world.* All science is cosmology... it is a fact that metaphysical ideas – and therefore philosophical ideas – have been of the greatest importance for cosmology. From Thales to Einstein, from ancient atomism to Descartes' speculation about matter, from the speculations of Gilbert and Newton and Leibniz and Boscovic about forces to those of Faraday and Einstein about fields of forces, metaphysical ideas have shown the way."[29]

In Popper's terms these theories remained metaphysical (i.e. speculative and unfalsifiable) until the innovation of technological means by which they could be tested. The demarcation between a metaphysical theory and a scientific theory, in these cases, is one of historical accident and yet the role that they played in the development of science was often significant prior to the innovation of the means to test them. By rejecting metaphysics, positivists have amputated a significant source of scientific speculation and have rejected the validity of the narrative aspect of scientific theorising. This narrative aspect, as a kind of story telling, is a cultural phenomenon that attempts to situate human beings within a world-picture. In all cultures, human beings have told stories about how the world came to be. This active social and psychological construction of narrative about reality is a part of what it is to be human. Scientists *qua* human beings are no different in this respect. Positivists have made a crucial error by rejecting this essential aspect of science as scientifically legitimate. The importance of intelligible explanations for the acceptance of scientific theories cannot be easily dismissed. Although it always remains possible that a scientific explanation may be replaced at some future date with another explanation, this still does not discredit the claim that physicists seek explanations. Even though it may well be an important part of any explanation that it is taken to be (at least tentatively) literally true it is not actually necessary that it is true for it to qualify as an explanation. The fact that an explanation is intelligible says nothing about its truth. Lies are a good example of this in everyday social life. After all, who would believe an unintelligible lie? Intelligibility is essential for any intentional deception to be effective. This is also the case for any honest attempt to explain a phenomenon because even though it is necessary that any explanation must be intelligible it may be true or false and still be intelligible. For example, nineteenth century physicists found the explanation of light as electromagnetic waves moving through aether to be intelligible despite the lack of evidence for it. Its potential to make the motion of light intelligible was independent of its truth. Of course, to some extent theories must "fit the facts" in order to be widely acceptable, but if a theory does not explain "the facts" then it is an unsatisfactory theory because physicists seek to explain the existence of such phenomena in terms of the same causal accounts they develop and use to design, build, operate, interpret, and modify the machine performances that are intentionally produced to model natural phenomena. The explanatory power of scientific model, theories, and laws to provide intelligible explanations of phenomena in the complex real world is central to the whole process of establishing the intelligibility of natural science. The fact that working scientists are unable to predict the behaviour of natural phenomena, such as lightning, does not mean that they are unable to explain lightning in terms of the laws of electromagnetism, thermodynamics, mechanics, and optics. Physicists remain content to consider the phenomenon of a lightning storm to be explained even though they are unable to predict the behaviour of such a complex natural phenomenon. In fact, the "complexity" of such phenomena is itself postulated to explain their inability to accurately predict weather in general. The intelligibility of any model is independent from its truth-status, the ability of physicists to bring transformative powers into the world does not logically support the truth-status of their causal accounts.

The history of physics has witnessed the establishment of numerous "empirical laws", such as Newton's Universal Law of Gravitation, Boyle's Law for Ideal Gases, Ohm's Law of Electrical Resistance, and Ryberg's Empirical Law for Spectral Lines. These laws only suggested that there was a simple relationship at work in complex phenomena, but they did not explain why it had that form. Newton found his own law quite unsatisfactory in this respect. Physicists, such as Leibniz, Boltzmann, and Bohr, were driven to find a theory, or model, that not only matched the "empirical laws" but also explained that model in terms of fundamental entities and causal mechanisms. It seems that causal explanation has been essential to the historical development of scientific theories.[30] Furthermore, given that the content of these explanations are in terms of mechanisms, or causal accounts, then these explanations can take a functional role in scientific reasoning during the construction of further experiments. This gives physics an exploratory and developmental trajectory. It was for this reason that Charles Sanders Peirce criticised the empiricist assumption that accurate measurement was the essence of science because, as he put it, in novel experimental work measurements "fall behind the accuracy of bank accounts" and the determination of physical constants was "about on a par with an upholsterer's measurements of carpets and curtains."[31] The motivation to provide causal explanations of the phenomena of the experienced world is irreducible to the practical technological interest in the deduction of predictions. It is an attempt to satisfy a basic human desire to explain and unify otherwise disparate experiences in terms of fundamental principles and causes. The purpose of measurement goes beyond testing the deductions of theories. It is a route by which hitherto unexpected novel mechanisms could be discovered. Instrumentalism, like positivism and empiricism, fails to recognise that an important goal of physics is to produce explanatory theories that describe the law-like behaviour of the causal agents that lead to the phenomena of experience. It seems that working scientists tend to take some of their theories very literally indeed. The aim of theorising is not just a matter of achieving empirical accuracy, or predictive success. It is also a matter of using technical causal accounts to explain how to design, build, operate, and interpret machines in the production and explanation of novel phenomena of experience. From its onset, it has limited itself to using mathematical theories to describe the law-like behaviour of machine performances. It tests the "empirical adequacy" of its mathematical theories against measurements produced using instruments designed by using mechanistic causal accounts. As such, physics is implicitly an attempt to explain measurements and interventions in terms of causal mechanism and, hence, is a metaphysical, technological, and an inherently realist enterprise proposed to discover the ontology of the causes of phenomena and it cannot be considered to be based upon an empiricist epistemology. Experimental physics is not an empirical science in van Fraassen's positivistic sense.

Bhaskar's transcendental argument

Bhaskar claimed that a scientific realist interpretation of experimental science is necessary because only a scientific realist interpretation can make experimental science intelligible as an activity. Bhaskar described this interpretation as a transcendental argument because it is based on the question "what must be true in order for 'x' to be possible?" where 'x' is taken to be some self-evident fact about existence. Such arguments are premised upon facts regarding the evident (or actual) and conclude that there is a "more fundamental something" that is a condition for the possibility of these evidences (or actualities). Bhaskar started from the premise that experimentation occurs in science and asked what must the world be like in order for this practice to be intelligible. His question was: what makes scientific experiments possible? His answer was a transcendental argument based on the following premises:

(i) Scientific experimentation exists,
(ii) Experiments are physical and not just mental,
(iii) Experiments involve causal interactions with the material world,
(iv) Causal interaction is only possible because we are embodied beings,
(v) As embodied beings, we are subject to the same laws that govern the material world.

He presupposed that there are laws that govern the material world. His conclusion was that the same laws that govern the material world govern experiments. That conclusion presupposed conceptions of what human beings are and what the world is comprised of. It is a statement of a realist interpretation of science and an expression of the spirit of the enterprise of experimental science. However, as Bhaskar pointed out, there may well be alternative transcendental arguments that explain the same thing differently and possibly better.[32] Bhaskar offered us an alternative as an example. His alternative transcendental realist argument for a realist interpretation of science ran along the lines:

(1) Science exists;
(2) Science discovers underlying mechanisms;
(3) If there were no underlying mechanisms then science would not be possible.

He concluded that there are underlying mechanisms. Of course, even if we accept premises (1) and (2) it does not follow that the "underlying mechanisms" discovered by science are *in fact* "natural mechanisms". Bhaskar's argument presupposed mechanical realism from the onset and, yet, it is a metaphysics that he does not address. Bhaskar's argument is a statement of allegiance. His assumption of the necessity of a realist interpretation of experimentation for the intelligibility of science is based upon an appeal to the "internal rationale" of experimental physicists. By taking this "internal rationale" of experimental science as the only intelligible rationale, Bhaskar has conflated intentionality with actuality. This is a fatal move for a realist argument. It must be a criterion for any realist position that, in any practice, the intentions of the practitioners could be at odds with the actuality of those practices. It must be possible for someone to think that they are doing one thing when, in truth, they are doing something else. Otherwise no one could ever be fallible. Furthermore, given a plurality of available interpretations of any set of practices, there is a sufficient degree of ambiguity for those practices to be taken as successful by the practitioners whilst being seen as otherwise by the uninitiated. It is possible that experimental scientists could intend to reveal natural mechanisms but only produce artificial mechanisms. It is only necessary, for the continuance of the practice of experimental science to be internally intelligible, that the practitioners interpret the artificial mechanisms that they produce to be natural mechanisms. We only need to show the process by which this interpretation is made and it is externally intelligible as well. We do not need to accept that any "internal rationale" is justified by the continuance of the practices. For example, many religions have been practised for thousands of years. Is it the only intelligible explanation of the existence and continuance of any particularly long-lived religion that it must be based upon truth? If we were to adopt Bhaskar's style of argumentation then we would have to accept that it was. After all, the devotee, no doubt, would claim that their material practices were based upon the truth of their beliefs regarding the significance of those practices. Rituals of purification, healing, exorcism, or for the dead would be a good example of material practices combined with beliefs about the worldly (as well as other-worldly) significance of these practices. Many religious practices are embodied theories that appeal to some kind of intransitive cosmic order and that acting in accordance with that order will bring worldly powers and material benefits. Many devotees claim that their practices are successful.

However, non-believers would readily claim that those practices were based upon cultural conventions, authorities, social power structures, and traditions, etc. We could argue that religion has maintained its existence through the maintenance of certain social structures, powers, and beliefs, and find it intelligible despite the "false consciousness" we have ascribed to its practitioners. We could equally argue this way about the conditions for the existence of experimental physics and make it intelligible. It depends upon the maintenance of certain social structures, powers, and beliefs. We do not need to accept the authority of either set of social structures, powers, and beliefs, to make either religion or physics intelligible. A similar argument could be made against Hacking's famous confession to be a realist about electrons because physicists claimed to spray them on mobidium spheres in the search for free quarks. It seems to me that Hacking should also be a realist about spirits because witch doctors claim to use them to heal the sick. After all, forms of shamanism have existed for much longer than experimental physics, and, many people claim to have been healed by spirits. By both Bhaskar and Hacking's standards, the endurance of a practice is a criterion for its truth and its continuance would be unintelligible if it was not successful. Bhaskar's transcendental arguments presume that experimental science is only intelligible as an enduring human activity if real causal mechanisms exist independently of science. It supposedly follows from the fact that science exists, endures, and is intelligible, that real causal mechanisms must exist independently from experimental science. One can imagine an analogous transcendental argument for a realist theory of shamanism:

(i)   Shamanism exists;
(ii) It aims to achieve knowledge of, and access to, a spirit world for purposes of healing the sick, exorcising evil spirits, etc.;
(iii) It would not be intelligible as an existent set of practices, if it persists and does not actually achieve what its practitioners claimed that it did;

Presumably we should conclude that shamanism must necessarily achieve knowledge of, and access to, a spirit world that exists independently of shamanic practices. However, this argument, like Bhaskar's "transcendental argument", is circular. It begs the question. It is a *petitio principii* because it presumes its conclusion: that the practitioners actually achieve what they intend to achieve.

However, even if we accept that it follows from Bhaskar's argument that real causal mechanisms exist and are discovered by experimental science, we still need additional work to determine whether they are natural or not. It does not follow from the fact that a generative mechanism exists that it is a natural mechanism. It requires a presumption of mechanical realism to make this leap. My argument in this book is that it is equally intelligible that such mechanisms are complex indices for particular interactions for the human interventions and machine performances that are created by experimentation. As I shall argue, this creative process is analysable as a non-linear technological process and mechanisms are understood as abstract indices for the ongoing

juxtaposition, interconnection, and unification of otherwise heterogeneous productive agents. Experimentation is a labour process that is
not possible without human participation and is not completely under
the control of the experimenters. I shall discuss the complicated
processes of constructing and interpreting machine performances in
experimental physics in detail throughout the rest of this book, but my
argument is straightforward. Production is not purely an individual
human activity. Individuals have no productive capacities whatsoever.
We must engage in disciplined practices with tools, machines, materials, and other people, in order to have any productive capacities at all.
We must relinquish any possibility of absolute control to become pro-
ductively empowered human beings. An experimental physicist does
not have absolute control of the processes by which s/he performs an
experiment. S/he must learn techniques and skills to become disciplined and be able to perform the technological practices required to
design, build, operate, and interpret experiments. S/he is empowered as
an agent by embodying these technological practices. Her/his intentions and actions are organised and directed through a technological
framework in which s/he is situated as an agent. This technological
framework has its own history of innovation and development.
Furthermore, due to the plurality of possibilities inherent to all experiments (otherwise there would not be an experiment), the challenge
for experimentation is to explore the productive possibilities of innovation. In order to be able to build, use, and interpret, experimental
apparatus, any human agent needs responsive to the pluralistic, incomplete, open-ended character of innovation. This involves relinquishing
absolute control over his/her agency during a participatory relation to
the performance of the apparatus. The human subject is empowered
and de-centred by this participatory relation in a context in which the
determination of the final possibilities of any experiment is perpetually
deferred to the future. Bhaskar claimed that any adequate philosophy
of science must grapple with "the central paradox of science" that
science is a social product that is concerned with the "knowledge of
things that are not produced by men at all."[33] It is my intention to cut
through the Gordian knot of this "paradox". My claim is that it is the
divergence of the centre of the technological framework from the individual human agent, and the dependence of this framework upon
human participation, which presents "the paradox" that the objects of
production are a social product and yet not completely controlled by
human beings. There is, in fact, no paradox at all if we do not assume
that the technological is simply "man made". Unless human agents

had thought that technological means could discover natural mechanisms then there would not be any experimental sciences. The objects produced through scientific activity depend upon scientific thought as a condition of their existence. However, as objects produced in technological contexts, their behaviours are not controlled by scientific thought. The behaviours of these objects are dependent upon and structured by the technological process of experimental science. There is nothing more or less mysterious about this than any creative act of making or material practice (e.g. composing a piece of music or inventing a novel dance). Experimental science is a creative labour process that involves the innovation, description, and explanation of machine performance in non-linear, changing, and complex contexts. This labour process depends on scientific thought as a condition of its existence and is, in part, characterised by scientific thought, but it is not controlled by scientific thought. Technological objects are made within the historical innovation and development of heterogeneous agents into unified technological frameworks. Experimental sciences are complex artificial processes and, therefore, any adequate philosophy of experimental science must grapple with the question: what is the artificial? Bhaskar has presumed that the answer to this question is simply "man made" and, consequently, he is confronted by a paradox that he can only resolve by assuming mechanical realism or considering experimental physics to be impossible.

Bhaskar's transcendental argument is committed to mechanical realism because he identified the necessary objects of experimentation to be "natural mechanisms" that exist independently of any experiment, whilst remaining capable of being produced within the closed system of the experiment. However, the general ontological conclusion that he derived from his interpretation of scientific practices as being rational and successful begs the question. It is the concept of the rationality of any scientific growth or change that is at stake for any scientific realist interpretation to get off the ground. An interpretation of physics is not inherently flawed if it questions, or even rejects, any notion of rational scientific inquiry. Anti-realist arguments cannot be criticised on the basis that they do not sustain a concept of scientific rationality. They are designed to undermine such a concept. The onus is upon scientific realists to provide such a concept because without it there is no rational basis for scientific realism. To criticise anti-realist interpretations of science because they do not provide the basis for a scientific realist interpretation is an unreasonable criticism. Bhaskar is not the only scientific realist who has made such an unreasonable criticism of anti-realist interpretations of experimental science. Christopher Norris, for example, based his whole critical realist argument against positivist and anti-realist interpretations of quantum mechanics on the premise that anti-realist interpretations of quantum theory must be flawed because they are not realist.[34] In my view, Norris missed the anti-realists' point about the limitations to what physicists can legitimately say on the basis of quantum physics if they are to use the results of those experiments to support their scientific epistemology. Furthermore, his argument begs the question. It would have been better if Norris had given an account of how measurements are made within quantum physics and shown that anti-realist interpretations of quantum theory, such as Bohr's interpretation, were premised upon a realist interpretation of the construction of the experimental apparatus that provided the experimental results that supported anti-realist interpretations. As I argued above, empiricist interpretations of experimental physics presuppose a realist interpretation of how and why the experiments were designed and built in the first place. Furthermore, the construction of experiments, such as the double-slit interference experiment, presupposes the existence of entities such as electrons, and even if one adopts a positivistic position regarding the ontological status of the measurable states of an electron, the fact that such an experiment is said to demonstrate wave-particle duality about electrons entails realist commitments. The Copenhagen Interpretation of this experiment not only presupposed that the quantum properties of such systems exist, but that they can be produced and observed utilising the properties of such devices as cathodes and phosphorescent screens. Although we can account for scientific change by locating the structures of change within the structures of social powers, once we take technology into account then we can go further than that. It is possible to provide an interpretation of rationality in experimental physics that allows technical growth and change, within the productive contexts in which technical choices and selections are made. A concept of "bounded technical rationality", as described in the previous chapter, sustains a notion of rational scientific growth and change, without requiring any commitment to the truth of scientific realism, providing that this concept is understood in the context of its mechanical realist heritage. This provision does not require any commitment to the truth of those precepts because we only need to address their function within the establishment of the template for subsequent scientific practices and its development. It does not matter whether these precepts are true or not. All that matters is how they function within the discursive and technological practices of experimental physicists. This does not require the existence of natural laws or mechanisms as anything more than abstractions of sets of discursive and technical indices for the dynamic interactions between technological objects.

Bhaskar's assertion that events are "categorically independent of experience", in order to reject the empirical definition of events in terms of experience, is necessary but insufficient for a realist theory of experimental physics. The realist needs to show that the means by which events and experiences are produced within the laboratory are also categorically independent. However, this "categorical independence" is not apparent in real experimental physics. It is blurred by the interplay between representations and interventions during the ongoing work of disclosing novel phenomena and attempting to make stable and communicable observational techniques. Of course, it does not immediately follow from that absence of categorically independent experience within experimentation that observation is subjective. An accepted technique of measurement is required and events are determined from within the development of the technological framework of the experiment to the extent that the experience of the phenomenon in question is constructed in terms of a set of measurements and responses. The phenomenon is defined by this set of measurements and responses. Thus events and experiences are brought together within the technological framework and they can only be taken as categorically independent by making the technology involved transparent. This is achieved by forgetting its history. Reference to an unobserved event (e.g. the ionization of a gas as a sub-atomic particle passes through it) is made via the connection between theoretical and technical causal accounts in reference to an observed event (e.g. the clicks of a Geiger counter). Bhaskar's realism treats these two kinds of reference as if they were causally connected because he assumes a tight and competent link between scientific observation and technical causal accounts. He has uncritically accepted a causal connection between the technical expertise and intentionality of scientists. So did Hacking. However, this assumption is inconsistent with both of their theories of science. If one accepts that our knowledge of the intransitive objects of experience is itself transitive, as Bhaskar does, then we cannot assume that there is a tight and competent link between technical causal accounts and interpretations of the event. If our interpretations change, which both Bhaskar and Hacking affirmed, and our skill at making interpretations improves, which both Bhaskar and Hacking asserted, then we cannot, at any stage, assume that our current interpretations are correct and, therefore, we must address the interplay between experience, events, and explanations.

When Bhaskar claimed that he has presented a basis for "rational principles of action" he implicitly offered us a pragmatic principle of action that is characteristic of bounded technical rationality directed towards the achievement of *techne*.[35] The constant conjunction of events produced in the closed system of experiment is only necessarily governed by natural law if we claim that such pragmatic principles are natural principles. Otherwise we have merely metaphorically substituted *techne* for "Natural Law". We could alternatively argue that physics achieves progress only by extending the variety technological objects at its disposal. Contemporary physicists have more techniques, materials, tools, machines, and instruments at their disposal than seventeenth century physicists did. Furthermore, the contemporary physicists have the recorded efforts of the previous generations of physicists at their disposal. In a technological context, contemporary physicists are able to deal, on an everyday basis, with far more complex, sophisticated, and powerful machines, instruments, techniques, and tools than the seventeenth century physicist would have been able to imagine. It does not immediately follow from this innovative productivity that the contemporary physicist has one more iota of knowledge about Nature than the seventeenth century physicist (or an ancient Greek, for that matter). This notion of "progress" in terms of an extension of technological powers does not provide necessary and sufficient reasons to presuppose that it does, in fact, discover any natural principles that exist *a priori* to the practices of physics. Thus experimental physics could be said to achieve "progress" only in a context of technological expansion, whilst its practitioners (and some philosophers of science) were under the sway of mechanical realism, but whether physics has progressed in epistemic knowledge is the very question at stake. Bhaskar did not establish any realist argument for a rational dynamic of change. He merely asserted that there is one. He cannot provide such an argument – or should he – because science is unfinished and we do not have its conclusions at our disposal. It has not achieved its own *techne* and, consequently, physicists do not know what it is that they have done and are doing, if we demand of them that they provide a complete causal account of all their experiments and their implications. Physics is experimental and its success is still open to question. After all, we cannot say that we are nearing the truth, improving our approximations of the truth, until we know what the truth looks like. We can only claim nearness if we assume that there is a final form of truth and then pre-empt it by assuming that the theoretical interpretation of the successful innovation of any technique to make novel technological power is a step nearer the final truth about Nature. The formula of "success equals nearness to truth" in the context of experimental physics is premised upon a conflation of knowledge with the acceptance of technique. Henceforth, within this formula, convergence with objective truth is equated with increased technological power. Bhaskar's transcendental argument has neglected the constructive role of technology in experimental physics and is based upon a false ontological dualism between human activity and Nature as the only possible poles of control at work in experimental activity. This leaves Bhaskar open to the following criticisms:

(i)   His transcendental argument for realism is circular and is mere assertion;
(ii) His realist interpretation of the rationale of experimental activity is based on the hidden presumption of mechanical realism and thus is an "internal" rationale;
(iii) If the constructive role of technology is taken into account then we do not have to assume that either experiments are purely human constructions or that they must reveal the laws of Nature that exist independently of human activity.

If we take technology into account then we see how the truth disclosed by experimental physics is neither purely a fiction nor necessarily reveals any natural laws, but is the product of a technoscience that creates and transforms the reality it reveals by creating and transforming itself. This insight opens up the possibility of interpreting experimental physics as a metaphysical performance art and reveals the implausibility of the positivistic demand that all metaphysics is purged from science. The positivistic claim that science should be free from metaphysical speculation should be rejected because all truly novel theories require metaphysical interpretations. New experiences require new concepts and, if we are to understand those experiences, we need to understand the metaphysics that made those experiences possible. In the next chapter I shall discuss the origins of this metaphysics and how it made these new experiences possible.

3
The Mathematical Projection of the Six Simple Machines

Historians and philosophers of science generally consider modern physics to be inherently mathematical and often cite Galileo's most famous quotation from *The Assayer* regarding the necessity of mathematics to read "the Book of Nature":

> Philosophy is written in this grand book, the universe, which stands continually open to our gaze. But the book cannot be understood unless one first learns to comprehend the language and read the letters in which it is composed. It is written in the language of mathematics, and its characters are triangles, circles, and other geometrical figures without which it is humanly impossible to understand a single world of it; without these, one wanders about in a dark labyrinth.[1]

In this chapter I shall address the question of how the self-evidence of this extraordinary statement was conceptually possible. How was the mathematical, empirical, and natural science of modern physics conceptually possible? In order to answer this question we need to examine the transcendental metaphysical foundation of modern experimental physics. What conception of the physical was presupposed in order to allow a mathematical and mechanistic science to be presented as a natural science? Heidegger analysed the essence of modern science in terms of the transformation of fundamental concepts that constituted the "scientific revolution" of the fifteenth to seventeenth centuries. What was the transformation of fundamental concepts that occurred during this period? How do modern sciences differ from ancient *episteme* and medieval *scientia*? How was this transformation possible? In his 1929 inaugural lecture *Was ist Metaphysik?*,

Heidegger considered modern science to be distinct from ancient *episteme* or medieval *scientia* because "in a way peculiar to it, it gives the matter itself explicitly and solely the first and last word."[2] What is the way peculiar to it? What is "the matter itself"? How does this way give "the matter itself" words at all? These questions are crucial for understanding Heidegger's analysis of modern science and its "special relation" to the world. He examined this special relation, and the human pursuit of science that guides and sustains it, by attending to how modern science relates to the world and what happened in order to attain that relation. In this pursuit, the human being, as one being among others, "irrupts" into the whole of beings in such a way that, in and through this "irruption", beings supposedly show themselves as what and how they are. What is the character of this "irruption"? How does it help beings "show themselves as what and how they are"? It was essential to Heidegger's approach that human beings are only able to pursue science by anticipating the nature of the being that they pursue. Human beings could not begin (or end) the pursuit without anticipating the conditions under which it could be considered to have been successful or have failed. How and what do human beings "anticipate" when they pursue science?

Heidegger accepted that measurement, experimentation, the use of mathematics, and relating conceptual and material practices, are characteristics of modern science, but argued that these characteristics should not be taken to be the essence of modern science.[3] He considered experimentation, as a means of acquiring information and testing cognition via a definite ordering of things and events, to be a basic kind of experience and activity involved in all craft work, tool use, and material practices. This was familiar to ancients and medievals alike. Ancient, medieval, and modern sciences involved working with mathematics and measurements, and were alike in this respect. The use of facts, experiments, measurements, and mathematics, was not the fundamental novelty of the emergence of the modern sciences and were not the fundamental characteristics of the "scientific revolution". Modern science is different from its predecessors because of *the way* that it measures, experiments, uses mathematics, and conceptualises. The metaphysical foundation of the mathematical aspect of physics was central to Heidegger's understanding of the fundamental event in the pursuit of modern science, in the work of Galileo, Descartes, and Newton. He termed this pursuit to be *mathematical projection*. For Heidegger, the meaning of mathematical projection did not derive from mathematics itself because mathematics is only a particular formulation of the mathematical. The word "mathematical" referred to the way that something is learnt, rather than merely using mathematics, and the word "projection" referred to the fundamental pre-suppositions and expectations that anticipated the phenomenon. Galileo, Descartes, and Newton conceived the motion of each and every body as having one "basic blueprint" according to which motion was nothing more than the determination of geometrical points in uniform space and time. This "basic blueprint" circumscribed its realm of application as both universal and uniform. The conception of a body moving under uniform and universal rectilinear motion, as posited by Galileo and Newton, was one that did not correspond to any experienced motion of a body and there is not any conceivable experiment that would bring such a body into direct perception. Heidegger noted the irony in the positivistic rejection of medieval scholasticism and *scientia*, as merely dialectical and poetic, in favour of a science concerned with an imaginary and non-empirical thing such as uniform and universal rectilinear motion. How could a law that describes an imaginary abstract, as a fundamental representation of things in contradiction to experience, found this new "empirical" science? Of course, it is arguable that we could travel into the vacuum of space and put Newton's First Law to direct experiential test. There are two problems with this argument. Firstly, Newton's First Law was widely accepted 250 years, or thereabouts, before our technological ability to perform this experiment and we cannot claim that experience was a criterion for its acceptance. Secondly, even if we were to perform this experiment, how we could *prove* that the body was *in fact* moving in a straight line?

For Heidegger, mathematical projection was a conceptual project of conceiving the essence of phenomena that skipped over the phenomena and opens a domain where facts can show themselves. He used the term "skip over" to focus on the way that modern mathematical physics does not actually attend to the phenomena. In my view, Heidegger intended a double meaning to this term. On one hand he intended the connotation of "brushing the phenomena aside" and on the other hand intended the connotation of rapidly (and lightly) stepping over appearances to reach "their essential reality". For both connotations, the notion of the phenomena as being an obstacle or a hindrance is implied. According to Heidegger, both Galileo and Newton started with an attendance to their experiences of the phenomenon of movement but "skipped over" it in order to conceive of natural motion in terms of an abstract uniform and universal rectilinear motion. This project posited that motion was to be properly conceived of in a specific way, and what and how it was to be evaluated was brought to the phenomenon. It was axiomatic and the cognition of motion that was taken and posited in the mathematical project was of such a kind as to set things upon their foundation in advance by expressing experience in terms of fundamental propositions. As axiomatic, mathematical projection anticipated the essence of things by sketching in advance the "basic blueprint" of the structure of everything and its relation to every other thing. This conception of Nature also required a mode of access and disclosure appropriate to its axiomatic predetermination. Which things were to be shown and how they were to be understood was prefigured in the project; the project determined the mode of experiencing and studying the phenomena. This basic plan provided the measure for laying out the circumscribed realm of Nature. The mathematical projection of uniform and universal motion would *in the future* determine which bodies could be a part of it and anchored in it. The project established a uniformity of all bodies according to relations of space, time, and motion; it also required, and made possible, a uniform measure as an essential determinant of things, i.e. numerical measurement. Things were determined within the "basic blueprint" only in terms of their positions in space and time, and as measures of mass and force. Henceforth, natural bodies could only be what they appeared to be within this projected plan. Inquiry was predetermined by the outline of the project in order to allow a line of questioning that posed conditions in advance to which Nature supposedly could only answer one way or another. Heidegger argued that, upon the basis of this project, experience became defined in terms of measurement and the modern experiment, and, consequently, modern science is experimental and numerical because of the mathematical projection of uniform and universal motion.

However, in order to appreciate fully the depth of Heidegger's insight, we need to carefully attend to how he defined mathematical projection. The mathematical projection of the "basic blueprint" used mathematics, in the narrow sense, but Heidegger meant much more than this. The new form of modern science did not arise because mathematics became an essential determinant but, on the contrary, the use of mathematics was a *consequence* of mathematical projection. The founding and application to natural philosophy of analytical geometry by Descartes, infinitesimal calculus by Newton, and differential calculus by Leibniz, were only possible because of the projection implicit from the onset. How are mathematics and the mathematical connected? Heidegger's definition of the mathematical began with its etymological stem in ancient Greek. He translated *ta mathemata* to be what can be learnt and taught, *mathanein* as to learn, and *mathesis* as the learning and the teaching. The two-fold meaning of *ta mathemata* was to teach and to learn in a broad and essential sense (and not the narrow and trite sense of schools and scholars). Heidegger understood *ta mathemata* in terms of his understanding of what is involved in truly learning something or truly teaching something. True learning does not occur by merely being instructed that something is the case. It occurs when the student learns for him or herself, in terms of his or her "own" experiences, what the teacher is offering. Learning is a form of taking, self-giving, and is experienced as one's own. It involves realising for oneself what is being taught. Teaching involves letting the students learn for themselves by bringing them to the point of learning by bringing to the fore what the students are already capable of learning in their own terms. This way of learning is determined by what is brought to bear by the learner upon the phenomenon in question. Heidegger defined the mathematical according to the following general characteristic: it takes and gives to itself cognisance of something as a cognisance that it already had and brought to the experience of learning. The mathematical had the original meaning of learning what one already knows. For Heidegger, number was the most familiar form of the mathematical because numbers are the closest to that which we recognise in things without deriving it from them. He used the number "3" as an example. One cannot teach children the number "3" merely by showing them three chairs, or three apples, or three cats, and instructing them to see the unifying cognition of three things. Children must recognise that for themselves. Number is the most familiar example of the mathematical because it is the most readily learnt and taught. Other things are simply more difficult for children to learn for themselves, and, consequently, more difficult to teach. Recognition of one's own reflection in a mirror, one's own mother, that other people have feelings, acceptance of one's own mortality, how to read and write, and many other things that are not related to mathematics, would also be examples of the mathematical. For Heidegger, the numerical was something mathematical in this sense.

Citing Galileo's famous (and perhaps mythical) experiment of dropping weights from the tower of Pisa, Heidegger argued that onlookers disagreed with Galileo's interpretation of the same phenomenon. They saw the weights hit the ground at slightly different times, whilst

Galileo triumphantly upheld his view that they hit the ground at the same time. Heidegger's conception of mathematical projection has considerable commonality with Kuhn and Feyerabend's interpretations of the "theory ladeness" of Galileo's experiments and observations. It was because Galileo conceived the motion of all bodies as rectilinear and uniform that he could conceive of acceleration as changing uniformly when an equal force affected it. This was how Galileo could conceive of the motion of a body thrown onto a horizontal and smooth plane as being uniform and perpetual if the plane was extended infinitely. Galileo used this thought experiment to present his conception of the motion of a body in such a way as to allow the reader to cognate it for him or herself. Galileo's *a priori* conception anticipated the universal nature of all bodies. No bodies are special, every place is like every other, no motion is special, and he defined every force only in terms of the change in motion it caused. In this sense, Galileo's discourses were the presentation of mathematical projection as a *mathesis* in which the reader was to learn how to see geometrically for him or herself how things truly were. In Galileo's work, the mathematical was used to reflect what Galileo wanted to learn from phenomena and, consequently, reflected his anticipations. Galileo's science was as abstract as the scholastic natural philosophies that he criticised for their abstractness and his universal conceptions pre-empted the "empirical" characterization of the phenomenon of movement in terms of universal motion.

What are the limits and justification of mathematical formalism in contrast to a demand for a science based upon experience? We all have the experiences of objects falling when dropped, the sunrise and sunset, and seeing the moon move across the night sky. The act of categorising these distinct experiences as manifestations of universal gravitation requires a non-empirical conceptualization of these phenomena. Novel research involves the fundamental creation and extension of new concepts rather than collecting *mere facts* because a fact is only what it is in the light of a fundamental conception. The creation of new concepts is characteristic of Kuhn's "revolutionary science". The positivistic attempt to distinguish modern sciences as primarily based on facts and experience is inadequate. Positivistic science is only capable of performing the work of "normal science" by presupposing and applying these new concepts to experience as givens. We can only appreciate how the application of mathematics to experience developed from the mathematical, argued Heidegger, if we grasp the meaning of the mathematical at a deeper level. For Heidegger, every kind of thinking was a consequence of a mode of historical Dasein and was a consequence of fundamental positions taken towards Being and towards the way that beings manifest as such, i.e. towards truth. What new fundamental position of Dasein showed itself in the rise of the dominance of the mathematical? For Heidegger, this new fundamental position was a spirit and formulation of freedom (against the Church, faith, and Aristotelian dogma) to have new experiences. In the mathematical project an obligation to the principles, demanded by the project itself, was developed and self-imposed. How did the project, according to its inner direction, drive towards an ascent to a metaphysical determination of experience? The modern conceptions of science, mathematics, and metaphysics sprang from the mathematical as a *mathesis*. This possibility of "seeing the truth for oneself" in terms of principles and axioms that were "self-evidently and intuitively true" founded the spirit of liberation from the past.

Galileo and the mathematical projection of the six simple machines

Historians of science and technology have provided numerous examples of ancient and medieval measurements, experiments, the use of mathematics to analyse natural phenomena, and conceptualised material practices, that discredit the "received wisdom" that these facets began during the "scientific revolution".[4] Their common argument is that the ancient and continuing medieval innovation of technologies and the fascination with mechanism contained the seeds of the sixteenth century development of mechanics and the new physics. The physical concepts used by Galileo, Descartes, and Newton had significant continuity with those used in ancient and medieval studies of mechanics. The most influential ancient sources for the science of mechanics were the works of Heron, Archimedes, and Pseudo-Aristotle, and the medieval modifications and criticisms of these works of became the points of departure for the mechanics of the sixteenth and seventeenth centuries. As well as the geometrical mechanics of Pseudo-Aristotle, Heron, and Archimedes, several ancient texts dealing with the geometrical and general principles of craft practices were available, translated, and studied throughout the medieval period. These included the works of Vitruvius, Heron, Frontius, Plindy, and Pappus. Vitruvius (c. 1BC) wrote *De Architectura* on the theory and practice of architecture and the large-scale management of craftsmen and labourers. Heron also wrote detailed works on surveying instruments (*Metrica*,

Dispotra, and *Catopica*) and practical engineering (*Pneumatics* and *Automatopoietike*), as well as his studies of mathematics, physics, and mechanics. Frontinus (c. 1AD) wrote *De Aquis* on the engineering and distribution of water supplies. Plindy's (c. 1AD) *Naturalis Historia* included sections on artifice and the mathematical treatment of mechanisms. Pappus (c. 4AD) wrote on mechanics to solve practical problems that inspired Galileo's solutions to the same problems. All of these works contained systematic collections of geometrical treatments, inventions, designs, experiences, and accounts of established practices. If we consider technology to be the *logos* (rationale, accounts, principles) of *techne* then all of these works are technological. They do not merely constitute collections of accounts of trial and error tinkering.

The sixteenth century science of mechanics had its origins in the mathematical treatments of mechanical devices and the culture of technological innovation that occurred from, at least, the thirteenth century onwards. As White argued, during the thirteenth century there emerged both generalised concepts of mechanical power and the view that Nature was "a vast reservoir of energies to be tapped and used according to human intentions."[5] This suggests that the abstract conception of Nature as a resource, a conception that Heidegger considered to be the essence of modern technology that distinguished it from ancient handicrafts, emerged 300 years before the "scientific revolution" and over 500 years before the nineteenth century "industrial revolution". White argued that between the thirteenth and sixteenth centuries there were widespread innovations in civic, military, and economic technologies, as Europe began its expansion of political, economic, and military powers. This expansion required more resources and the continuing innovation of machines and techniques to enhance productive, explorative, military, and civic power. In his view, the sixteenth century development of technologies was a continuation of the post-thirteenth century period of decisive development in the effort to use the forces of Nature mechanically for human purposes. The invention of the printing press accelerated the cultural dissemination of the mechanical arts throughout Europe, whilst they were rationalised and associated with the ancients through geometry in order to give them status of being "sciences".[6] The know-how of craftsmen and practitioners was presented in terms of mathematical and rational principles and transformed into "true knowledge". When coupled with the patronage of political and military powers this elevated the social status of the mechanical arts and prepared the way for the experimental and mechanical "natural" philosophies of the six-teenth century to achieve mechanical leverage into the workings of Nature. The experimental and mechanical sciences of the sixteenth and seventeenth centuries grew out of the contemporary mechanical arts and mathematical sciences of the fourteenth and fifteenth centuries. The artisan had a central role in the emergence of experimental philosophy because, as Bennett put it, the "experimental philosophy, given its methodology of testing hypotheses by manipulating mechanical devices, must be said to have appropriated both values and specific knowledge from the mechanical arts."[7] After transforming the status of mechanical invention from a craft to a science, via mathematics and the patronage of elites, the conditions were ripe for the formal construction of the science of mechanics.

The predominant mathematical problem for medieval mechanics was how to solve the problem of the six simple machines (the wheel and axle, the wedge, the balance, the lever, the inclined plane, and the screw). The methodological template to provide complete solutions for these machines followed the Aristotelian mechanics set down in Psuedo-Aristotle's *Mechanical Problems*.[8] In this work, the formulation of the geometrical solution to the six simple machines utilises the properties of circles. Since the mid-thirteenth century, many treatises appeared that focussed on the problems of kinematics and dynamics for mathematical and philosophical treatment. The thirteenth century mathematician Jordanus de Nemore, in his book *De rationale ponderibus*, used Pseudo-Aristotle's dynamics, as well as Arabic derivatives, to tackle the problems of the balance, weights, and levers. His work focussed upon the problem of geometrically dealing with motion. By the fourteenth century, there were many books on the subject of the application of geometry to the problems of motion. John Buridan's fourteenth century book *Questiones super libris quattuor de caelo* included discussions of impetus theory, the possible rotation and motion of the Earth, the general law of leverage, the solution to the problem of the inclined plane, and the equilibrium of connected weights.[9] Both Nemore and Buridan tackled problems that were to become central to Galileo's work. Of course, we might raise the objection that these early efforts utilised completely different conceptions of motion and matter. Buridan's notion of "impetus", for example, has no correlate in Galileo's mechanics and, in terms of modern mechanics, the early efforts made significant errors in their treatments of even simple mechanical devices. However, this is irrelevant for the question of whether Galileo's mechanics was a culmination of earlier efforts. What we see in the pre-sixteenth century efforts was *the attempt* to describe the motions of simple mechanical devices in terms of Euclidean geometry. This set down the template for the methodology of subsequent efforts.

Medieval mechanics within this tradition attempted to construct a dynamical system based upon the geometrical projection of the circle. However, this system failed to provide a method to provide static solutions to the problem of equilibrium. One of the most valuable sources for this project was the works of Archimedes. Roger Bacon, in the thirteenth century, invoked Archimedes against those who did not "dare to know" and Commandino, in sixteenth century, wrote, "with respect to geometry no one of sound mind could deny that Archimedes was some God."[10] Archimedes' fame as an inventor of fantastic machines was widespread in the fifteenth century, largely through the account of Plutarch's *Life of Marcellus*.[11] It is ironical that it is claimed in this text that Archimedes destroyed all his designs for machines because of the ignobility and danger of such records.[12] Although famous for his mechanical inventions there are no surviving texts, directly attributed to Archimedes, which contain his reputed devices; only second and third-hand accounts remain. His surviving works include the geometrical solutions to the sphere, cylinder, conoids, spheroids, the equilibrium of planes, spirals, buoyancy, quadrature of the parabola, the diameter of the Earth, numbers, square roots, irrational numbers, arithmetic, a method of integral calculus, the diameter of the Universe, probability, solids, centre of gravity, and measurements.[13] There are references to lost works on polyhedra, numbers, balances and levers, gravity, optics, the mechanical motions of heavenly bodies, parallel lines, circles, triangles, and machines. His only remaining description of a mechanical device is his orrey to model the mechanical motion of the heavens. In the *Method* (addressed to Eratostheses), Archimedes described "a certain method, by which it is possible for you to get a start to enable you to investigate some of the problems in mathematics by means of mechanics" and wrote "for certain things first became clear to me by a mechanical method, although they had to be demonstrated by geometry afterwards because their investigation by the said method did not furnish an actual demonstration."[14] This mechanical method had wide appeal to the medieval and Renaissance mechanists and mathematicians.[15] Leonardo Da Vinci studied both Archimedes' mathematics and Psuedo-Aristotle's mechanics and regarded those who did not read Nature "by the light of experience" with contempt.[16] His works included mathematical analyses of machines, in terms of mathematical mechanics, which were reduced, primarily, to the elements of force, impact, weight, and motion. This analysis integrated geometry and mechanical arts – an integration of both Archimedean and Aristotelian theories of geometrical mechanics with fifteenth century technological practices – and postulated that the principles of mechanics and the principles of Nature had an analogous explanatory connection. However, he regarded "experience" not in terms of just observation, but in terms of an interventional exploration of "the processes of Nature" through chemistry, mechanics, and dissection. These "processes" were treated in his experiments, models, and art, as mechanisms.[17] Leonardo's approach of exploring natural phenomena in terms of "natural mechanisms" experimentally, applying geometry to solve problems in natural philosophy, was a continuation of medieval efforts in these directions rather than a radical or novel break from them.

It is beyond doubt that the works of Archimedes had a profound influence on Galileo. This profound influence can be read from Galileo's own words: "I cover myself with the protecting wings of the superhuman Archimedes, whose name I never mention without a feeling of awe."[18] In 1586 Galileo constructed a hydrostatic balance, following Archimedes' geometrical arguments, to determine accurately the relative amounts of two metals in an alloy mixture, which he described (in Italian) in a paper published in 1644.[19] In 1586 Galileo also studied the Archimedean concept of "the centre of gravity" and wrote a paper (in Latin) on "Theorems about the Centre of Gravity in Solids".[20] Galileo wrote *On Motion* (*Du Motu*) applying Archimedes' principle of motion in a medium whilst retaining Aristotle's notion of natural places and the medieval notion of impetus in 1590–1.[21] He argued that by treating a falling body as a body rising, falling, or floating in a medium, then, in Archimedean terms, such a body was "reduced to weights of a balance."[22] He wrote *On Mechanics* in 1593–4 giving Archimedean geometrical treatments of the simple machines.[23] He started from his premise that all simple machines could be reduced to a problem of an Archimedean balance. This argument was based on the Archimedean principle that all machines operate on the same physical principles so a complete understanding of any one of them is adequate for the deduction of the mechanical properties of all the others. Having chosen the balance as fundamental and used it to derive the laws for an inclined plane, the lever, the windlass, the capstan, the pulley, and the screw, Galileo constructed a "dynamic equilibrium" method as the basis of his physics. He used this method in his treatment of hydrostatic phenomena in *Discourse on Floating Bodies* (1612) and in his *Dialogue on the Two Chief World* Systems (1632). He used this method to describe motion as separated into two independent horizontal and vertical axes to describe the fall of a body from a moving point as that of a parabola. He *rhetorically* argued that the Earth could revolve around the Sun (without the breath being snatched from our mouths nor birds being flung from out of the sky). In his last work, *Two New Sciences* (1638), he concentrated on explaining natural motion using the inclined plane.[24] This involved using the pendulum experiment and the balance as exemplars for the description of all natural motion.

Galileo probably first became acquainted with the works of Archimedes in 1583 through the Tuscan court tutor, Ostilio Ricci, a pupil of Nicolò Tartaglia (1500–57), who, in 1543, had translated the works of Archimedes into Latin.[25] Tartaglia taught perspective, architecture, and in 1537 published his mathematical science of ballistics.[26] He also taught mathematics, surveyed land, designed fortifications, made maps, and invented mathematical instruments. His studies included arithmetic, geometry, music, astronomy, perspective, and architecture. He had translated Euclid's works into Italian in 1543 and, in 1551, he published his Italian translation of Archimedes' *On Floating Bodies* and, using Archimedes' hydrostatics, Tartaglia derived and proposed a method of re-floating wrecks.[27] He had also studied and translated Pseudo-Aristotle's works into Italian and declared that mechanics based on the principles of weight was the cause of every ingenious mechanical invention.[28] He argued that arguments about Nature could only be based on experience whereas abstract arguments about mechanics should be based on mathematics. This lead him to assert that arguments based on mechanics were superior to those based on mere observation because reasoning based on mathematics was more rigorous than reasoning based on experience. When observation and mechanics did not agree then the notions of "error" or "material hindrances" could be used to explain the discrepancy.[29] Margaret Osler, following Mandelbaum, termed this as the methodological problem of *transdiction*.[30] This form of inference constructs an explanation in terms of an unobserved mechanism in order to explain the deviation of an observation from theoretical expectations. This method was central to the use of mechanics as an explanatory tactic and, by adopting this tactic, Tartaglia had pre-empted the mechanical philosophers of the seventeenth century. Galileo frequently used this tactic. The discrepancies between the path of a cannonball and a parabola, for example, could be transduced as the mechanical consequence of the invisible force of friction. A subsequent mechanical experiment could be constructed to demonstrate friction and, due to the presumed universality of such a demonstration, it could be taken to have disclosed the reason for the initial failure of mathematical description to match experience. Tartaglia accepted the Aristotelian classification of mechanics as a "subalternated science" because its method was abstract mathematical demonstration but its subject was physical and consequently both mathematics and experience were required in the development of mechanics. He argued that mechanics provides knowledge to calculate the strength (*virtù*) and power (*potentia*) of any machine to augment the strength and power of men by any degree.[31] Tartaglia attempted to inscribe a formal mathematical treatment of mechanics by combining the statics of Archimedes with the dynamics and kinematics of Psuedo-Aristotle. He was unsuccessful because he could not consistently combine the Archimedean proofs based on equilibrium and the Aristotelian arguments based on velocities. However, Tartaglia had laid down the challenge to his sixteenth century Italian contemporaries.

Francesco Maurolico was the first of Tartaglia's contemporaries to take up this challenge.[32] He already had established his reputation in astronomy, optics, and by translating and commenting on the works of Euclid, Archimedes, and Pseudo-Aristotle. In *Problemata mechanica cum appendice* he discussed the scope and classification of mechanics within the sciences.[33] He listed mechanics (along with music, astronomy, perspective, geography, architecture, painting, sculpture, stereometry, and cosmography) as an intermediate science between the mathematical and the physical that was distinct from the secular arts. He considered mechanics to be a part of "contemplative philosophy" due to its mathematical part. He argued that the dynamics and kinematics of Pseudo-Aristotle's mechanics had to be based on "the doctrine of equal static moments" and consequently mechanics had to be based on Archimedean principles. It is this notion of the primacy of Archimedean statics that inspired many subsequent sixteenth century Italian mechanists. For example, Guidobaldo, a sixteenth century Italian aristocrat and a military engineer, used Archimedean techniques to solve the problems set by Psuedo-Aristotle starting from the lever and then on to the rest of the six simple machines.[34] He aimed to establish mechanics as a branch of rigorous axiomatic geometry and claimed that any machine based on such a mechanics would work in the real world. Another sixteenth century Italian military engineer called Giulio Savorgan, also inspired by Archimedes, innovated Italian town fortifications, developed mechanics

and invented "Archimedean" devices to aid the lifting and transport-
ation of heavy cannons.[35] He invented light, robust, and powerful
lifting-gear based on spur gears, worm gears, rack-and-pinion, block-
and-tackle, winch and pulley, screw jacks and ratchet-jacks. The
Aristotelian mathematician, Bernardino Baldi considered mechanics
to be a "subalternated science" due to its physical subject matter described in terms of geometrical proofs.[36] In his view, mechanics was
consequently of an equal status to optics, music, and astronomy. In
his treatment of mechanics he followed Maurolico and formulated
his solutions to the problems raised by Pseudo-Aristotle in terms of
Archimedean and Aristotelian concepts of statics, dynamics, equilibrium, motion, power, and impetus.

The Aristotelian and mathematical science of mechanics was established in Italy through the influence of the university at Padua.[37] The
elevation and establishment of mechanics from the banausic to the
academic occurred at Padua through the influence of mathematically
educated Aristotelian scholars such as Niccolò Lenico Tomeo and
Alessandro Piccolomini.[38] Since the fourteenth century Padua had been
a centre for mathematical subjects (including astronomy, astrology,
geometry, optics, and geography) and was the first Italian University in
the sixteenth century to offer lectures in mechanics from the chair of
mathematics. The introduction of mechanics into the curriculum at
Padua in the 1560s began in the form of lectures on Pseudo-Aristotle's
work. Guidobaldo studied there in 1564 and Baldi from 1573 to 1575.
Pietro Catena was the first lecturer in mechanics at Padua and gave lectures between 1564–1573 and was professor of mathematics from 1547
to 1576. His successor was Guiseppe Moletti.[39] Moletti followed his
Aristotelian predecessors and classified mechanics as "contemplative
philosophy" of mathematical principles of statics, dynamics, and kinematics. According to Moletti, the task of mechanics was to demonstrate
the most efficient means of performing the maximum amount of work
with the minimum of effort. For Moletti, mechanics was a science and
not an art because the geometrical first principles of mechanics were
"necessary and eternal" whereas the arts were contingent upon human
ends. The end of science was the knowledge of causes and truth
whereas the end of arts was productive work. He argued that the first
principles of mechanics were natural means, that mechanics was to be
found in all the works of Nature, and the first principles were "Natural
Laws". Moletti transformed the traditional classification of mechanics
from a "subalternate science" to a "natural science". Whilst he still
considered it to be "intermediate" between the geometrical and the

physical it was based on both mathematical and natural principles and truths.[40] As professor of mathematics at Padua between 1577 and 1588, he paved the way for Galileo's mechanical realist physics. In 1592, Galileo succeeded Moletti as professor of mathematics at Padua.

Francis Bacon independently presented a similar natural philosophy in *The New Organon*.[41] He was critical of sixteenth century arts, intellectual sciences, and philosophy and reserved his praise for mechanics. He considered Greek science to be childish, due to their basis on "bland and specious generalities" that lead only to "disputes and scrappy controversies" and "almost stopped in their tracks", and praised the mechanical arts for progressing.[42] He proposed

> the production of a Natural History by making a history not only of Nature free and unconstrained (when nature goes its own way *and does its own work*), such as a history of the bodies of heaven and the sky, of land and sea, of minerals, plants and animals; but much more of *nature constrained and harassed* when it is forced from its own condition *by art and human agency, and pressured and moulded*. And therefore we give a full description of all the experiments of the applied part of the liberal arts, and all the experiments of several practical arts which have not yet formed a specific art of their own.[43]

He argued that the natural axioms induced from experience founded the mechanical arts. He praised the mechanical arts for providing a "variety of objects and splendid equipment", having "contributed to human civilization", and being based on "axioms of nature" discovered by observation and subtle, patient, ordered movement of hands and tools. If directed according to utility they were capable of growth and flourishing. He cited the clock as an example of "a subtle and precise thing that seems to imitate the celestial bodies in its wheels, and the heartbeat of animals in its constant, ordered motion; and yet it depends on just one or two axioms of nature."[44] He considered the mechanical arts to be praiseworthy as the source of civilization and political advantage in general and the discovery of the art of printing, gunpowder, and the nautical compass in particular. The mechanical arts were the noblest human pursuit and "right reason and sound religion would govern its use."[45] As Bacon put it,

> There remains one hope of salvation, one way to good health: that the entire work of the mind be started over again; and from the very

start the mind should not be left to itself, but be constantly controlled; and the business done (if I may put it this way) by machines.[46]

Another proponent of this ontological interpretation of mechanics at the turn of the seventeenth century was Giovanni di Guevara.[47] Guevara was a Spanish noble from Naples, *praepositus generalis* of the Clerics Regular Minor, the Bishop of Teano in 1627, and a papal legate to Philip IV of Spain. In his *In Aristotelis mechanicas commentarii*, published in 1627, he analysed mechanics using both Archimedean principles and Pseudo-Aristotelian mechanics. He dealt with the principles of mechanics, centres of gravity, the simple machines, Psuedo-Aristotle's thirty-five mechanical problems, the scope of mechanics, and its relation with the other sciences. He defined mechanics as the art or science of applying geometrical principles to heavy and light things that must be moved or brought to rest artificially. Mechanics was based on the weight of the moved body and the strength of the mover (which could be an impetus or a machine) and it consisted in discovering the appropriate powers needed to move loads and to supplement Nature. Guevara's formal treatment of mechanics was in terms of marvellous and artificial motion and rest (each, in turn, was treated in the Aristotelian terms of violent and natural motion and rest). In the Aristotelian terms, violent motion arose from an external source whereas natural motion arose from the body in question. These distinctions allowed him to describe how mechanics and natural philosophy dealt with the same subject differently. He argued that natural philosophy was concerned with marvellous motions and rest whereas mechanics was concerned with artificial motions and rest. Both natural philosophy and mechanics could analyse their distinct kinds of motion and rest in terms of natural and violent motion. He argued that natural motion was apparent in any motion that was produced by machines. Although a machine operated upon violent motions, from an external source, the behaviour of that motion could be analysed in terms of natural motions. In the operation of any machine there were both violent and natural motions at work. In other words, human intervention was required to produce and activate any machine and Nature played its part in how that machine operated. Consequently, he based the understanding of the operation of mechanical devices on the interaction of human interventions and natural principles.

These novel interpretations of mechanics provides the key to understanding Galileo's "natural philosophy" by placing the methodology that checked his conclusions against "experience" within the context of mathematical and practical mechanics. Indeed, it is true that he pre-supposed that "mere" experience did not qualify as a legitimate source of knowledge. He frequently disregarded the primacy of perception and affirmed Euclidean geometry, inscribing motion solely in terms of the translation of a body from one geometrical point to another. Checking directly with Nature involved an intellectual *a priori* knowledge of mathematics, as being the only means of apprehending the truth about experience, and physics became *a priori* science in the hands of Galileo. Geometrical methods did not allow any investigation into quality. It is for this reason he declared and dismissed the notion of quality as illusionary. His "insight" was that only by describing experi-ence in the form of Euclidean geometry one was able to read "the book of Nature". However, his science posited that Nature is comprised of mathematical properties that can be isolated and treated as interacting components in a mechanical apparatus. His geometrically arguments were projected to disclose "natural mechanisms" by analogy with demonstrations using artificial devices. For Galileo, demonstrating these deductions of mathematically described mechanisms involved practical experimentation because if one truly understood any natural phenomenon then one should be able to construct a machine to repro-duce that phenomenon artificially.[48] Galileo's method to "check" his theories against "experience" was a technological process of using abstract mathematical designs to demonstrate the possibility of build-ing mechanical devices to "imitate" natural phenomena. Success implied that he had "reproduced" the causal mechanisms supposedly "at work in Nature" when he designed a machine that reproduced the appearance of natural phenomena. Hence, he rhetorically claimed that his mathematical deductions disclosed "natural mechanisms" when those deductions were successfully "embodied" in design of a working machine. This is evident from Galileo's use of a pendulum to demon-strate his theory of motion, his use of an astronomical sphere to demonstrate his theory about the Sun's rotation, and steelyards and balances to demonstrate his theory of free fall.[49] Mechanical realism was implicit to his methodology. This was an essential development for Galileo's new physics and provided the template for all subsequent experimental physics. The mechanisms at work in the construction of mechanical artefacts were *literally* those of Nature. Mechanics was an embodiment of mathematics in the world. In the metaphysics of Galileo, mechanical realism had emerged into seventeenth century as a substantive metaphysical position and constituted a set of precepts.

Mathematical science required this set of precepts in order to appeal to a generalised principle of operation in Nature in order to correlate the motion of bodies, and their properties, with measurements and experiments.[50] Galileo's mechanical realism restricted the classification of "the real" to be "the mechanical" and the mathematical motions projected upon the natural phenomena were the abstracted motions of the six simple machines. The geometrical solution to the wheel projected circular motion and the coupling of anti-parallel linear motions. Orthogonal changes in motion, the transference between horizontal motion and vertical motion, could be described in terms of the geometrical solution to the wedge (the transference between vertical and horizontal described in terms of the inclined plane). The projection of the screw abstracted transference between circular motion and motions orthogonal to the plane of the circle. Galileo had reduced all these machines to the operation of the lever and the balance.[51] This provided him with a complete set of uniform mechanical motions to mathematically project on to all natural movements in such a way as to allow the observer to "see" the geometrical essence of that motion.

This *mathesis* founded his arguments that the efficient causes of mechanics were the necessary causes and fundamental mechanisms of Nature. It was the balance that, as a metaphor and a model, that was to become central to physical explanation and law.[52] The subsequent development of the laws of conservation (mass, charge, energy), Newton's Laws of Motion, and the First Law of Thermodynamics, all were premised upon the metaphor of the balance as a fundamental mechanical principle of Nature. Furthermore, by utilising the method of transdiction whenever the mathematical projection of the balance failed to match experience, the projected invisible counter-mechanisms can be used to correct the discrepancy by projecting it as a counter-balance. An example of this in Galileo's work is the solution to the discrepancy of the motion of a body from the mathematically projected quadrature of a parabola. Galileo explained this discrepancy in terms of air resistance.[53] He demonstrated the explanatory power of this transdiction by using the dropped weights and the pendulum thought experiments. Thus natural processes and change could be represented as the resultant interaction between balances, balances within balances, and the lever. Mathematically projecting the lever over any change in motion defined that change in terms of an external force. Galileo was able to develop the dynamics of the new physics that aimed to describe everything in terms of number, figure, motion, and causal mechanism. Hence, for Galileo, the precepts of his metaphysical mechanical realism encapsulated his conception of Nature and the methodology of natural science.

Galileo made essential two contributions to mechanics that made modern experimental physics possible. The first contribution was technical. Galileo innovated geometrical techniques to reduce all motion to a single unitary mechanical motion: the motion of the balance and, therefore, he was able to inscribe simple time-reversible mechanisms, such as pendulae, in terms of Euclidean geometry and provide a mechanical determination of time. The second was metaphysical. He was able to rhetorically establish his mechanical realism as a basis for using mechanical devices, experimental apparatus, to "discover" mechanical principles of Nature. Galileo went further than had any of his Paduan predecessors. Not only were the motions of simple mechanical devices treated as natural, as Moletti had proposed, but they were also to be used to determine the mathematical "Laws of Nature". It was the geometrical treatments of simple mechanisms that were to be classified as "the Laws of Nature" and all natural movements were to be treated as simple mechanisms. Galileo's reductive mechanical realism was both the precursor to the "mechanical world view" and provided a method to investigate Nature mechanically. The mathematical projection of balances and levers made the mechanical world-view and the idea of the clockwork universe possible. All that was required to solve the problem of transdiction was to devise a further experiment to show the mechanical action of the correcting mechanism. Mathematically projecting this transdiction over the original phenomenon refined the model and provided an explanation. This projection embodied the precepts of mechanical realism and made the experimental use of mechanical devices to ascertain the fundamental mechanisms of Nature possible. The subsequent speculative metaphysics of the seventeenth century, such as the Newtonian mechanical system of the world, Gassendi's atomism, or Descartes' metaphysics, constituted the epistemological novelty by introducing interpretations that would have been nonsensical to previous natural philosophers. Differing speculative metaphysical assumptions founded the subsequent mechanical philosophies, but they all presupposed the operational metaphysical precepts of Galileo's mechanical realism.

The mechanical realism inherent to Galileo's development of mechanics as both a mathematical and a natural science conflated *techne* and *episteme*. He presented the *techne* of mathematical mechanics as the unchanging and eternal *episteme* of Nature. Thus, from its onset, mathematical physics was *techne*-logical. This move was facilitated by the ambiguity between mathematical reasoning as *techne* and *episteme* within the texts of both Aristotle and Plato, but it was also novel. This novelty allowed both the transformation of the status of the mechanical arts to mechanical science and the reification of the products of the mathematical science of mechanics. Henceforth, mathematics became the means of reading "the book of Nature". However, this transformation occurred within the context of the European desire for novel technological powers. It is this desire that provided the condition for naturalness of the conceptual synthesis of the precepts of mechanical realism and the possibility of both mathematical natural science and modern scientific technology. Securing this metaphysical foundation connected the *episteme* of Euclidean geometry with the *techne* of mechanics as a technological means to disclose the axioms of Nature that governed change in the world. The technical acts of writing the book of Nature could be ignored (as mere means) and it could be read as if written by God. The mathematical practices of Galileo's physics allowed the intuitive *mathesis* of the skilled practitioner to become a means to discover the truth. This remained the case in European mathematics until the nineteenth and twentieth centuries' construction of non-Euclidean geometries. The precepts of mechanical realism, embodied in mathematical projection of the six simple machines, allowed both modern experimental physics and modern technology to be conceptually possible. Mechanical realism underpinned the sixteenth and seventeenth centuries' conception of technological powers and natural phenomena as having the same unitary origin and manifest according to the same principles or laws. The conceptual possibility of modern technology occurred simultaneously with the possibility of modern experimental sciences because of the emergence of this conceptual synthesis of the origins of both natural phenomena and technological powers. Once this symmetry emerged then the conception of modern technology as a unitary phenomenon, manifest according to universal natural laws, was possible. This was simultaneously a naturalization of mechanisms and a mechanization of Nature. The conception of technology as a process of unlocking and utilising natural forces, causes, and powers became possible. It could be treated as a unitary kind of relationship between "Man" and "Nature", in which "Nature" could provide the means for its own domination by "Man". Henceforth, technology was a neutral process that was accessible to "universal rationality", defined in terms of "technical rationality" in accordance with a concept of "efficiency". Henceforth, Nature was conceptualised in terms of universal mathematical laws, materials, mechanisms, necessity, and efficient causes. This premise allowed technological innovation (bringing novel technological powers into the world) to be taken as nothing more than taking advantage of those laws. Technological innovation (the expansion of technological powers) could be treated as human participation in the natural order of things disclosed by the mathematical mechanical sciences. In the sixteenth and seventeenth centuries, the confidence in the human ability to construct and use machines to produce new technological powers grew to such proportions that it found expression in the reification of machines. The presupposition of mechanical realism allowed these reified machines to become transparent means of disclosure at the service of "Man". Machines could be used to disclose the mechanical principles of the Grand Machine, the Universe and everything contained therein, and conceived as a self-evidently rational exploration of Nature. Appeals to technological innovation and power underwrote the validity of the new science.

Mechanical realism and the mechanical world-view

Heidegger argued that the new science, by setting in place each particular being in such a way that calculation provided certainty of the objective reality of that being, transformed the conception of Nature into an object of explanatory representation. Henceforth, scientists could only consider "the calculable" to be real within scientific research. Modern scientific research was only possible when and only when successful calculation provided the "test" of the truth of any hypothesis or theory. Heidegger argued that this conception of truth and the philosophical establishment of mathematical projection, as the definition of modern metaphysics, began in the work of Descartes (as articulated in *Meditationes de prima philosophia*). The usual interpretation of Descartes' *cogito sum* is that of the thinking being, "I", as the human subject. This self-declared centre of thought placed doubting at the beginning of philosophy in order to provide reflection upon knowledge itself and its possibility, placing epistemology before ontology. Heidegger criticised the usual interpretation that Descartes' philosophical project was a form of scepticism, subjectivism, or egoism. For Heidegger, this interpretation was a story that at best is "only a bad novel" because it neglected the questioning of substance that was central to Descartes' philosophical enterprise, as articulated in the *Meditationes de prima philosophia* (1641).[54] Heidegger placed Descartes' work within the context of an historical period in which a new assault upon tradition had begun that was to be "seen in [the] fact that man frees himself from the bonds of the Middle Ages in freeing himself to himself".[55] Descartes' enterprise reflected the passion for this new assault and an inquiry to bring clarification to the essence of the new enterprise. This enterprise not only emphasised that clear and insightful intuition, or certain deductions, are the routes to knowledge, but also held that method, in general, is necessary for us to have truths at all. This method was to consist in the order and arrangement upon that which "the sharp vision of the mind" is to be directed if truth is to be discovered. If mathematics, in the sense of *mathesis universalis*, was to ground and form the whole of knowledge, then special axioms were required. These axioms needed to be intuitively self-evident and establish in advance what constitutes being and from where, and how, the essence of being is to be determined. The basic mathematical projection had to be based upon its own ground, as a basic principle, and be indubitable. Descartes did not start his discourse with doubt because he was a sceptic, but in order to clear the way for positing the mathematical as the absolute ground and foundation of natural philosophy. Hence Descartes' *cogito* was something mathematical, in Heidegger's sense, when in thinking itself it takes cognisance of itself as something we realise for ourselves as something we already have. Descartes' formula, *cogito ergo sum*, was not an inference because the *sum* was the *fundamentum* rather than the consequence of thinking. This formula founded the "I posit" proposition because it allowed his work to be presented as something independent from whatever is given before hand and as that which already lies within. This can be clearly seen in his posthumous work *Regulae ad directionem ingenii*. Descartes articulated the aim of his life's work to be the project of grounding the mathematical in terms of its own inner requirements by explicating it as the standard of all thought and establishing its rules.[56] Descartes' self-appointed task was a work of reflection upon the fundamental meaning of the mathematical. This reflection was concerned with the totality of beings and the knowledge of that totality, and, therefore, was a reflection upon metaphysics. The mathematical "I" was presented as the special subject, before whom all remaining things first present themselves as what they are, that mathematically provided the fundamental relation from which all things receive their thinghood. In relation to the "subject", things could then stand as something else, as *objectum*, and be "objects".

However, Heidegger neglected to attend to the influence of the sixteenth century science of mechanics upon Galileo and the mechanical realist metaphysics that his new science assumed. He failed to recognise that Descartes was profoundly inspired by Galileo's work in his definition of the methodology for the new science and, consequently, the same mechanical realist precepts underpinned his metaphysical foundation to natural philosophy. As we can see from a letter that Descartes wrote to Florimond de Beaune (dated April, 1639), he considered the new physics as "merely mechanics".[57] In *Meditationes de prima philosophia*, he conceived all natural phenomena in terms of machines and asserted that

> there are absolutely no judgements in Mechanics which do not also pertain to Physics, of which Mechanics is a part or type: and it is as natural for a clock, composed of wheels of a certain kind, to indicate the hours, as for a tree, grown from a certain kind of seed, to produce a certain kind of fruit. Accordingly, just as when those who are accustomed to considering automata know the use of some machine and see some of its parts, they easily conjecture from this how the other parts which they do not see are made: so, from the perceptible effects and parts of natural bodies, I have attempted to investigate the nature of their causes and of their imperceptible parts.[58]

Descartes was committed to Galileo's mechanical physics and natural philosophy, and, consequently, his aim was to establish a mechanical philosophy of Nature upon the foundation that all natural phenomena could be explained in terms of innate matter and motion in geometrical space. Physics was the project of the mathematical projection of the six simple machines over all natural motion and change. He argued that the Universe is a plenum and that the matter filling it is infinitely divisible, identical with geometrical space, and has only the property of extension. He argued that extension could be understood in terms of *a prior* knowledge, there is no need for any appeal to experience or observation, and consequently, the first principles of natural philosophy could be known *a priori* and lead to the discovery of essences. Experiences and observations were only required to determine the contingent actuality of phenomena. His *Meditationes de prima philosophia* contained his demonstrations of the metaphysical foundations of the epistemological basis of his mechanical philosophy that would replace the Aristotelian natural philosophy without appealing to an alternative ancient philosophy.[59] Not only did he use sceptical arguments instrumentally in order to clear the way for his arguments in favour for mathematics as a foundation of indubitably certain and demonstrative knowledge, but he also used theological arguments for God choosing to be bound by the necessity that God had freely created in the physical world. Consequently, Descartes' *a priori* arguments for the eternality, universality, and necessity of God's choice of mathematical first principles were the metaphysical basis of his epistemology of physics.[60] Furthermore, his points of departure were continuous with the theological presuppositions of medieval theology.[61] His metaphysical arguments about God's creation of eternal, necessary, and universal truths, were situated within traditional discussions about the absolute and ordained powers of God, and these played a formative role in the development and interpretation of Descartes' natural mechanical philosophy. Descartes, Galileo, and Kepler identified mathematical truths as eternal truths that were central to the natural order of the physical world and all shared a concern with the relationship between God and mathematical truths.[62] Descartes' arguments for God's creation of mathematical truths provided the metaphysical foundation of his epistemology because if certain fundamental mathematical truths are necessarily true then we could have *a prior* knowledge of them. These *a priori* indubitable truths provided Descartes with a foundation for his deductive methodology. From this "standard of certainty", Descartes was able to provide arguments in the *Discourse on Method* for his method of systematic doubting, the *cogito*, the existence of God, the existence of the soul, and the essence of matter. He was able to instrumentally start from his conception of the *cogito* based on the components of doubting, thinking, and being, to argue the *cogito* was indubitable as a transferable standard by which the reliability of any knowledge claim could be made. This standard provided "a general rule that the things we conceive very clearly and very distinctly are all true, but that there is nevertheless some difficulty in being able to recognise for certain which are the things we see distinctly."[63] Descartes argued that if this general rule were true, and it must be, then there is a necessary connection between that which is clear and distinct in our minds and the natural order of the physical world created by God. If the rule were false God would be a deceiver and this would be in contradiction with the conception of God in terms of perfection.

Given that mathematical truths are clear and distinct then they must provide truths of the physical world. Descartes used this reasoning to establish his characterization of matter in terms of geometrical extension, infinite divisibility, and primary and secondary qualities. These characterizations constituted the fundamental elements of the physical world within Descartes' natural philosophy. God was the first cause of motion and always conserved an equal quantity of motion in the Universe according to laws of inertia and impact. Descartes appealed to the perfection of God in order to justify the possibility *a priori* knowledge of these laws of Nature. His derivation of the existence and content of these laws from God's attributes required knowledge of God's attributes. As a consequence of the perfection of God, the same laws of Nature would govern any world created by God, and therefore in order to obtain knowledge of this particular world more than just the *a priori* laws of Nature are required. For Descartes, knowledge of the laws of Nature was necessary but not sufficient to explain particular phenomena. He required observation and experiment to explain the phenomena of Nature because we needed to know which of all the possible phenomena are existent in this world and which of the several possible mechanisms, compatible with the same general law, governed the production of the phenomena in question. Even though the laws of Nature were eternal and necessary, the actuality of the phenomenal world was contingent because the particular implementation of the laws of Nature was contingent.[64] To know which mechanisms God had used to make the phenomenon in question, as well as which phenomena God had made, one needed to observe and experiment. In Descartes' discourse, the laws of Nature represent the possibilities of God's choices of phenomena and mechanisms when making the actual. In terms of the possibility of human knowledge, observation and experiment were constrained in terms of what could be made or manipulated mechanically either in practice or in thought. Descartes' scientific method was to produce mechanical analogies (or models) derived from first principles that would produce the same phenomena observed to exist in the world. Observations and experiments could then be used to eliminate deduced mechanical models from the potentially infinite set and provide criteria by which judgements regarding which mechanisms were the actual mechanisms involved in the production of the phenomenon in question. By using "empirical evidence" to eliminate deduced possibilities, except one, Descartes hoped that the demonstrative character of his natural philosophy would be secured. Experiments and observations were not designed to validate laws of Nature but rather to select from a set of possibilities and to show how the general laws applied to particular phenomena. The purpose of experiments and observation were not to provide data for the induction of general laws. Descartes' scientific method was to deduce possible mechanisms from *a priori* laws of Nature as proposed explanatory mechanisms potentially eliminated via observation and experiment. These laws were used to deduce the possible explanatory mechanisms and the purpose of empirical investigation was to eliminate all the unexercised possibilities. He proposed that a mechanism, or set of mechanisms, was at work in producing the phenomena and machines could be used to determine the truth of any explanation by attempting to produce phenomena *artificially.* An understanding of the mathematically rationalised arts, transformed into mechanical sciences, constituted the basis for an understanding of the productive capabilities of God. Furthermore, once we understood these productive capabilities then we too could become more God-like in our capacity to change and produce things in the physical world. Hence, Descartes wrote,

> we can have useful knowledge by which, cognisant of the force and actions of fire, water, air, the stars, the heavens and all the other bodies which surround us – knowing them as distinctly as we know the various crafts of the artisan – we may be able to apply them in the same fashion to every use to which they are suited, and thus make ourselves masters and possessors of Nature.[65]

Descartes' natural philosophy intimately bound together the human capacity to make with the discovery and subsequent implementation of natural mechanisms. By securing epistemological validity to productive success, Descartes was able to secure the knowledge of natural principles to productive skills. Henceforth, he secured the validity of a theory in terms of the innovation of novel technological powers associated with its application, and argued that any failure to implement his theories in productive practices was possibly a failure on the part of the craftsman rather than a failure in the theory.[66] Newton used similar arguments regarding his prisms.[67] If a prism failed to resolve the seven-coloured spectrum then Newton would argue that the craftsman lacked the skills to make it properly. Newton's definition of a good prism was that it showed a seven-coloured spectrum. Henceforth, natural philosophers and scientists were able to make good rhetorical advantage out of the problem of auxiliary hypothesis. If we cannot know whether the primary hypothesis was in error, or an auxiliary hypothesis used in the deduction of observable possibilities from primary hypothesis or the construction of the experiment to test that primary hypothesis, then there is not any logical "test" of theory at all. The failure to implement a theory in practice can always rhetorically explained away by using the method of transdiction, criticising the theory, or by criticising the experiment.

It is important to situate Descartes within social context. In the early seventeenth century, an influential group of self-professed mechanical philosophers emerged. These people established a community of writers dedicated to the establishment of the metaphysical foundations of mechanical philosophy, the promotion of the growth of the new mechanical sciences, and the opposition to Aristotelians and the occult. As well as Descartes, the members of this community included Beekman, Cavendish, Charleton, Digby, Gassendi, Hobbes, and Mersenne.[68] These men corresponded with each other, reacted to each other's work, and formed an international intellectual community. This community had a formative influence on the next generation of mechanical philosophers, such as Newton, Boyle, Leibniz, Pascal, Huygens, and Hooke.[69] All of the "natural" philosophies produced by these individuals were implicitly premised upon the operational precepts of mechanical realism and mathematically projected the abstract motion of the six simple machines over all physical change. This mathematical projection underwrote the subsequent speculative interpretations of reality regarding contingency and necessity, the nature of matter, cause, and the ontology of the world. Mechanical realism had allowed the seventeenth century experimental and mechanical philosophies to be possible. It was then the task of the natural philosophers to build their speculative metaphysics upon those precepts. Their disagreements focussed upon concerns about which speculative metaphysics provided the most intelligible account of mechanical Nature and squared with their theological commitments. The mechanical world-view was both a speculative and operational metaphysical world-view. Its explicit assumptions about the constitution of the world allowed physical inquiry based on mechanics to be possible. These assumptions were emergent during the developing understanding of, fascination with, and confidence in the possibilities and potentials of machines. The metaphysics of mechanical philosophy was reduced, in accordance with the limits of the mechanization of processes, in such a way as to allow machines to have the power of disclosing natural mechanisms at work. This was possible because the conceptions of Nature had themselves been reduced to that of mechanical processes. In other words, the fundamental principles of Nature were reduced to be the fundamental principles of mechanics and consequently mechanics could be presented as the means by which the fundamental principles of Nature could be discovered. The circle was completed. Physics was the project of the mathematical projection of the six simple machines over all natural motion and change.

It was this new philosophical movement that began through the studies of mechanics, anatomy, and astronomy, during antiquity and the Medieval, and finally emerged in the sixteenth and seventeenth centuries as "new sciences". The (albeit limited) successes of the two mathematical sciences of astronomy and mechanics inspired the mechanical philosophers to propose that the motions of the entire physical world could be completely described in terms of laws, mechanisms, and mathematics. The physical world was to be described as nothing more than inanimate matter in motion in geometrical space – exactly the same components that comprised the conceptual basis of the rationalization of mechanical devices. Newtonian natural philosophy became possible. Newton was able to assert that mechanics should not be limited to the manual arts, but, instead, be used to investigate "the forces of Nature" and to deduce the motions of the planets, the comets, the moon, and the sea.[70] In Newton's *Principia* we can find a statement of his precepts of presented as the "Rules of Reasoning in Philosophy."[71] The first rule is an epistemological statement of the natural economy of causes that demands that we limit our investigation of the causes of natural things to the causes we identify as necessary and sufficient to explain the appearance on natural phenomena. Newton presumed that simplicity is a natural principle and consequently simple explanations were the most likely to be ontologically true. The second rule is a statement the invariance and universality of cause-effect sequences and, consequently, we should assign the same causes to the same effects. Newton's assumption that Nature is both isotropic and homogeneous is an essential assumption for all experimental physics because without it the experimenter could not extend the particularities of any local experiment to the universal level of a law. The third rule is a statement of methodological reductionism that allowed the universalization of the properties of bodies from those properties identified through experimentation. This assumption was necessary for Newton to assert that "the qualities of bodies are only known to us by experiment" whilst simultaneously allowing the qualities determined through experimentation to be informative about bodies upon which an experiment has not been performed. These rules display considerable continuity with Galileo and Descartes' precepts. In order to understand their connection with mechanical realism we need to examine their context of application. Newton founded geometry upon mechanics and told us that it

> is nothing but that part of universal mechanics which accurately proposes and demonstrates the art of measuring. But since the manual arts are chiefly employed in the moving of bodies, it happens that geometry is commonly referred to their magnitude, and mechanics their motion. In this sense rational mechanics will be the science of motions resulting from any force whatsoever, and of the forces required to produce any motions, accurately proposed and demonstrated.[72]

For Newton, all causes of motion were mechanical causes and, consequently, his rules of reasoning are a statement of the precepts of mechanical realism. Hence the mechanisms disclosed through the mathematical projection of machines could be taken to be both the universal mechanisms of Nature and as methodologically available for transdiction.

Similar views can also be found in the works of Robert Boyle and Robert Hooke. Boyle sought to explain cold, heat, magnetism, and all other natural phenomena in terms of mechanical principles. For example, Boyle wrote

> That which I chiefly aim at, is to make probable to you by experiments, that almost all sorts of qualities... may be produced mechanically; I mean by such corporeal agents as do not appear either to work otherwise than by virtue of the motion, size, figure, and contrivance of their own parts (which attributes I call the mechanical affections of matter).[73]

This is also evident in the case of his development of the air pump as a means to disclose the fundamental nature of (already presumed) homogeneous and isotropic space as a vacuum (or void). Once the technological innovation of this device was established (transformed into a reliable technological means of disclosure) then subsequent innovations and modifications could be woven into the social fabric of material science. This weaving was rhetorically secured to Boyle's natural philosophy via the public success in establishing the air pump as a repeatable technological device. The truth of theory was henceforth deferred to its future implementation in productive practice. The knowledge obtained from constructing and performing experiments was itself provisional on its use in the construction and operation of future experiments.[74] The epistemological criterion for any knowledge claim was that it could be instrumentally functional in the subsequent innovation of further machines. This epistemological criterion became central to the whole methodological enterprise of experimental physics. Hooke described his "natural philosophy" as the real, the mechanical, and the experimental philosophy.[75] He advocated the transformation of "natural philosophy" from the observational, experiential, and categorical, into a process of making interventions into natural entities and processes using instruments and machines to produce explanations of the sensible phenomena of experience in terms of fundamental mechanical interactions. Hooke's experimental mechanical philosophy was premised upon an intimate relationship between mathematics, natural philosophy, and machines. Hooke frequently used machines to present illustrations of "the common rules of mechanical motions" that he assumed were the mechanical principles of Nature. Newton, Boyle, and Hooke's natural philosophies represented machines as having explanatory power about Nature. They were able to assert the dream of deriving the rest of the phenomena of Nature from the same kind of reasoning from mechanical principles.

Once the status of mechanics had been transformed from a banausic art to a natural science, by the mathematical projection of the six simple machines as geometrical demonstrations, then those first principles could be presented as "eternal and necessary truths". In combination with the mechanical realist metaphysical premise that "natural causes" were efficient, this transformation allowed mechanics to be naturalised. The distinction between the artificial and the natural was dissolved for particular aspects of technology: the fundamental principles of mechanical motion. That was subsequently taken as self-evidently true and there was not need of any further metaphysical argument. Once this had been achieved then the ontology of experimental physics, based on mechanical apparatus, could achieve an epistemological legitimacy as a means of disclosing truth. Mechanical realism had become techno-ontological: it was a means of disclosing the truth and nature of beings. The "scientific revolution" of the sixteenth century was the mechanists' revolution that was founded upon the establishment of mechanics as a mathematical science and was directed towards the establishment of the epistemological legitimacy of mechanics as a natural science. The establishment of this legitimacy involved a transformation of the conceptions of matter, cause, natural necessity, and the dissemination of the mechanical world-view, in parallel with rhetorical appeals to the practical successes of mechanics. The reduction of the lived-world to the mechanical world required a distinction between primary and secondary qualities, a distinction between those properties possessed by material bodies and the effects due to the interaction between material bodies with human sense organs and minds, to account for the fact that human experience is not of a mechanical world (an accounting that was itself a transdiction) and also provided the possibility of a mechanical account of human perception. It required a fundamental transformation in the conception of the human body. With the increasing interest in the development of the mechanical sciences in seventeenth century Europe, for the purposes of enhancing technological powers, the discourses of mechanical natural philosophers become dominant. Once this occurred then the path was cleared for the notion of "mechanism" to become the dominant explanative trope. This monolithic explanatory strategy was symptomatic of the accelerated mechanization of European social organization towards the monolithic goal of achieving technological advantages for the competing European social elites. This transformation was a profound shift from the contemplative scholarly logic and poetics of Aristotelian natural philosophy towards the construction of mathematically rationalised machines and novel technological powers. Contemplative and scholastic Aristotelianism had become obsolete and irrelevant.[76] Henceforth, technological innovation led the discovery of truth and modern experimental physics is directed by the "how does it work?" question. By directing research towards the identifications of the "workings" of the causes of the phenomenon in question, modern experimental physics equates "the real" with "the mechanism". Thus modern experimental physics requires a tri-partite ontology: (1) what is moved (the object); (2) what moves it (the mechanism); (3) what governs or describes that movement (the law). By presupposing mechanical realism, modern physics operates upon a conception of the unity of its object (Nature), a unity of its means (the methodology), and, consequently, is able to present itself as a unified science aiming to disclose natural laws. The ontology of the part of the world presented by modern physics as Nature, the complex of machines, has only extended itself. The tri-partite ontology of physics has remained invariant in its structure throughout this extension, and its content only varies according to which particular type of machine (with its associated mechanisms and laws) is under investigation.

This ontology reveals the extent that *episteme* has been transformed by modern physics in order to be presented as a *techneic answer* in terms of general and abstract causal principles which take the form of mathematically abstracted mechanisms. *Techne* as a directional principle and episteme as a directional principle have converged. *Techne* has become naturalised and *episteme* has become mechanised. The single point of distinction between these two is that the former has the experimenter as the efficient cause and the latter has "the inner workings of Nature" as the efficient cause. The work of modern physics is to generate a praxis that removes the experimenter from the account. What is left is then taken to be non-human Nature. This work can be done because nonhuman Nature has been constructed as the mathematically abstracted workings of nonhuman machines. This is the methodological work of mechanical realism. The machines themselves, through mathematical abstraction, have been removed from the account and the transformed *techne* is presented as *episteme*. This metaphorical substitution allowed two important desires to be offered a source of satisfaction. The first was the possibility of a comprehensible world picture of the world, human beings, and how human beings are situated in the world. The second was the promise of novel experiences and novel powers as the fruit of human labour and skill at making. The mathematical science of mechanics offered the second; the mechanical world-view promised the first. Furthermore, the conflation between *techne* and *episteme* in experimental physics is the transformation of the ideals of the human character: Bacon's dream for the human character was that of a rational material agent whose primary function was to labour, and whose reward would be new powers and new challenges for labour. The emergence of experimental physics, as a continuation of the ancient and medieval mathematical treatments of mechanics, was situated, from its onset, within a cultural desire for organised and innovative technological powers. The wide acceptance of the new science did not occur because of the predictive success of Copernicus and Kepler's mathematical treatments of planetary motions. There is not any possibility of experimenting upon planetary motions (at least not yet) and the predictive successes of Copernicus' system over the Ptolemaic system were very much exaggerated.[77] Experimental physics was primarily accepted because of its successful association with the innovation of mechanical devices within societies that valued the economic, political, and military advantages of technological innovations. Appeals to the successes of mathematical astronomers were *rhetorically* connected to the successes of mechanists, as part of the movement towards a unitary conception of natural science. This occurred within the cultural context of patrons for which the mechanists were providers of technological innovations for economic, political, and military goals.[78] The precepts of the mechanical realist metaphysics were required (at least implicitly) to conceptually connect, via the mathematical science of mechanics, the astronomical phenomenon of planetary motion with terrestrial mechanical devices, and present both as aspects of a unitary natural science. The mechanical science of Galileo was the culmination of Medieval and Renaissance developments of geometrical mechanics and technological innovations; it was not the radical break from his predecessors that it has been presupposed to be. However, his physics presumed and attempted to justify the precepts of mechanical realism, and it was this presumption and attempt that was novel, heralded the "mechanical world view" of seventeenth century mechanical and experimental philosophies, and was a prerequisite for conceptions of modern scientific technology and modern experimental physics.

4
The "Making" of the Ground-Plan of Nature

Thomas Kuhn argued that two distinct "traditions", the mathematical and the empirical, emerged from the mechanical philosophies of the seventeenth century.[1] Margaret Osler, following Kuhn, argued that these two "traditions" exemplified and manifested themselves in terms of two distinct styles of scientific practices governed by distinct metaphysical and epistemological assumptions.[2] Osler termed these styles as "conceptual frameworks" which differed in the emphasis that they placed on empirical evidence and mathematics in their interpretations of natural phenomena. She argued that they emerged from two distinct theological traditions, exemplified by in Gassendi and Descartes' natural philosophies, in the latter part of the seventeenth century. Gassendi argued that all natural phenomena could be explained in terms of atoms of inanimate matter and their motion in geometrical space. This speculative metaphysics, based upon the ancient atomism of Epicurus and Lucretius, postulated that the Universe is composed of atoms and the void. Gassendi argued that atoms possessed the qualities of size, shape, and heaviness, and consequently cannot be described in terms of *a priori* knowledge. His theory of natural philosophy was based upon measurement and also the assumption that essences were knowable only to an absolutely free God. On the other hand, for Descartes, size, shape, and location, were the primary qualities and, since these were all geometrical properties, then the essence of a material object could be known through mathematical reasoning. God, being perfect, was unchanging, and therefore, the mathematical laws of the Universe, created by God, were themselves perfect and unchanging. It was not a question of whether God could change these perfect laws. Once God had created the laws of Nature then God would not change them. It was for this reason that Osler considered the development of mathematical and empirical traditions to be a consequence of the Descartes vs. Gassendi theological arguments on God's relation to natural laws and necessity.

I agree with both Kuhn and Osler that the "empiricists" tended towards practical problem solving, whilst the "rationalists" tended towards the development of grand theories. However, mechanical realism was presupposed by both approaches and, in my view, both Osler and Kuhn missed the unitary essence that connected these two dimensions. Mechanical realism had been used to justify the reduction of the experienced world into mechanically accessible properties that could be mathematically projected over parts of the world and presented as the whole world. Once mechanical realism had become established (which it certainly had become by the latter part of seventeenth century) then, both mathematics and observation had become integrated through technology into experimental and mechanical natural philosophy. Experimental apparatus, as means of disclosure, were based on the interaction between mathematics and mechanics. The content of both mathematics and experience had been transformed by the mechanical realist precepts and the techniques utilised to disclose "natural mechanisms". Empirical evidence was restricted to variables and quantities that in principle could be measured using mathematically rationalised and calibrated technological devices (scientific instruments), even though it was not necessary that they were actually measured in practice. Mathematical treatments were limited to forms that could be both abstracted from mechanical devices and used instrumentally in the design, building, operation, and interpretation of such devices. These two conceptual frameworks were derived from the same mechanical realist metaphysical precepts and, as such, were the two dimensions of the same technological framework and did not constitute distinct metaphysical positions. The mathematical dimension was more apparent in the grandiose mechanical realists, such as Descartes, Galileo, and Newton. Their problems involved developing a mathematical description of the entire Universe, whereas the practical dimension was more apparent in the modest mechanical realists, such as Boyle, Pascal, and Newcommen, whose efforts were directed towards developing particular machines in order to solve particular problems. The degree of difference in attitudes regarding the new science and its possibilities constituted distinct teleological positings within the same technological framework. As such, this difference is indicative of a spectrum of dispositions regarding the question of what could be achieved with the new physics rather than necessarily constituting distinct formally assumed metaphysical positions. Without a formal metaphysics, dispositions do not form conceptual frameworks about the world, even though they may well be formally transformed into them, but, rather, constitute different tendencies towards acting within the world. It is for this reason that I agree with Hacking that the word "style" is not that helpful in characterising the difference between "rationalism" and "empiricism".[3] In my view, "priority" constitutes a clearer term for discussions of the distinction between practical problem solving and grand cosmological theorising. The term "style" should be reserved for the writing and presentation practices, which have been developed to convey scientific narratives.[4]

Due to influence of the science of mechanics on the mathematical and technological template of the new physics, neither priority could be placed under experimental test in terms of the other. Any experiments devised to facilitate such a decision would require the very assumptions that were being "tested". Both priorities were based on mechanics, assumed mechanical realism, and consequently there was not any mechanical means, either in deed or thought, by which a decision could be made as to the superiority of the one approach over the other. Both the so-called empirical and the mathematical traditions were dimensions of the same technological framework in such a way as to centre the distinction between the priority of the usefulness of mathematics to experiment or of experiment to mathematics. Either way, the central constraint was that of the mechanization of any hypothesis or proposition. It is this two-fold dimensionality that is central to experimental physics. These two dimensions are evident in Newton's approach in *Principia*. Newton was an "empiricist" in so far as he argued that all facts should be induced from experiment and re-evaluated in the light of further experiments. He was also a "rationalist" in so far as he argued that the demonstration of any truth should be deduced from mathematical first principles. It is also evident in his *Opticks*, where the lens is itself reduced to an optical lever that mechanically operates upon (otherwise) rectilinear rays of light. This is also evident in treatments of the phenomenon of polarization, which is treated in terms of the wheel and the lever template. It is evident in the eighteenth and nineteenth century studies of mechanics, optics, thermodynamics, and electromagnetism. How can mechanical machines, optical machines, thermodynamic machines, and electromagnetic machines be characterised in such a way as to reveal a general principle by which practical experiments and mathematical theories can be linked and shown to be manifestations of the same technological framework? It is my argument that this general principle was the methodological princi-ple of mathematically projecting the abstracted motions of the six simple machines over all subsequent kinds of machine. This is done during the innovation of those machines as stable and repeatable disclosures of mechanisms whilst maintaining their distinction in relation to the kinds of materials from which those machines were built. These distinct areas of physics all operated by reducing the motions of their respective objects to that of circles, anti-parallels, orthogonal reflections, levers, screws, and push-pulls. However, these distinct areas of physics cannot be reduced to one another because they involve distinct sets of stabilised mech-anical processes using distinct kinds of materials. The metaphysics of mechanical realism allowed machines, as the concrete object of research, to become invisible as a means of disclosing natural mech-anisms. It is when those disclosed "natural mechanisms" have been utilised in future productive activity then they considered to be "tested". Modern scientific research achieves progress by successfully implementing explanatory theories within the technological frame-work of the innovation of new kinds of machine and technological power.

Setting-up the ground plan

Heidegger's analysis of the characteristics of the modern age held "the world picture" (*Weltbild*) to be central.[5] In this context, Heidegger used the word "world" to refer to "what is, in its entirety" and used the word "picture", not in the sense of a copy or imitation, but, rather, in the sense of the colloquial expression "get the picture", to capture the way that we grasp the matter in question. "The world" in this case, in place before us, as a representation, and all that belongs to it and stands together in it, is a system, in such a way that we are acquainted with it as something that we are equipped and prepared to deal with. Thus "the world picture" was presented as that which we are prepared for and which we intend to bring and set in place before us as some-thing conceivable (graspable). This setting in place, representing, of the world involves an essential decision regarding what is, in its entirety. It is an anticipatory act of mathematical projection. It was this setting in place, as something objectively before us and at our disposal, that was, for Heidegger, characteristic of the modern age. There was not an ancient or medieval "world picture" that was transformed into the modern; having a "world picture" at all is characteristic of the modern age. This "picture" was produced in such a way that it represents, gathers, and orders to us an image of the world that affords us the position of articulating, securing, and organising a "world-view". This allows us to measure and draw up the guidelines for everything that is in accordance with our power for the planned calculation and manipulation of all things. It was this picturing which provides a graspable picture of the world in which the self can be situated as the one who grasps and science "is one of the pathways upon which the modern age races towards fulfilment of its essence, with a velocity unknown to its participants."[6] As such, in my view, the mechanical realist metaphysics of modern science and modern technology underpinned this foundation for the modern age and is a central participant in the set-up and trajectory of modern culture. However, this still leaves us with the question of how a "world-picture" connected theoretical and material practices. How was the "world-picture" mathematically projected over phenomena, in such a way as to make them graspable and calculable in advance, in terms of mechanical principles? How are theories and experiences connected within the technological framework of experimentation?

Modern experiments are technologically sophisticated projects involving a wide range of techniques, practices, machines, tools, tacit skills, and knowledge. The objects experimented upon, such as "electromagnetic fields", "paraelectric materials", "photons", "nuclei", "quasi-particles", "electrons", "quark-antiquark events", "superfluids", etc., require techniques and machines for their production, observation, and manipulation. Without those techniques and machines we would not be "aware" of these objects at all. The relationship between scientific experience and these "invisible" objects occurs by transforming the macroscopic objects of everyday experience into a means of disclosure. Physicists are concerned with macroscopic objects, such as machines, because these technological objects disclose the underlying causal mechanisms in operation in those machines. The object of scientific inquiry is not the machines themselves but, rather, the *techno-phenomena* that are produced by those machines. The properties of superfluids, the dynamics of phonons in crystals, the thermal capacities of metals, the properties of lasers, superconducting materials, solar neutrinos, the polarization of light, etc., are all complex objects which are only disclosed through the mediation of machines, theories, and techniques. The establishment of scientific facts and theories about such objects requires putting techniques to work. The machines, theories, and techniques put to work to make investigation of these objects possible, mediate experiences of these objects. The observational aspect of experimental work involves the active technical use (and modification) of theories, methods, and techniques. It is complex and there is not any possibility of being able to disentangle theories, techniques, and observations, except in hindsight through reconstruction. Take He-3 for example. Experimenters in Ultra-Low Temperature Quantum Physics, for example, do not directly experiment upon "natural" helium. He-3 is an isotope of the element helium and is itself a product of technological processes. It is the by-product of the nuclear weapons industry and (to a lesser extent) the oil production industry. What the physicists experiment upon is a purified sample of He-3 extracted from these industrial processes. By what standard is "purity" defined here? "Purity" is defined in terms of an established technique of purification and, in order to know whether a sample is pure or not, the experimenters must do so in relation to that technique. This anticipates what the essential properties of He-3 are and that which passes through the template will be defined as "pure" He-3 in accordance with its functionality. Thus helium is transformed from a "natural substance" to a technological product, its properties defined via techniques and machines, and it is the functionality of He-3 that is experimented upon. This functionality is defined in terms of "natural properties" because the purification technologies have been naturalised as a means of disclosure. In addition, the properties of He-3 disclosed by the experimental apparatus used by ultra-low temperature physicists are only those that can be disclosed using experimental techniques such as dilution refrigeration and voltage resonance, for example.[7] Anything else will remain unobserved. The observed responses of He-3 to the interventions of the experiment are experienced as the set of techno-phenomena of the theoretically interpreted interactions between the apparatus and the technological product He-3. Furthermore, He-3 is studied by physicists for the sake of understanding the quantum properties of superfluidity at ultra-low temperatures and is transformed in a set of techno-phenomena for use in the technological framework of the experiment as a means of disclosing these quantum thermodynamic properties. They are understood by the physicists to be realizations of the transfactual quantum mechanisms that are independent from He-3 and are otherwise swamped by impurities and higher energy interactions. He-3 is used as a technological object to disclose these subtle mechanisms because it is taken to be the bounded technically rational choice on basis of its functionality within the technologies at the physicists' disposal.

The scientific experience of any techno-phenomenon is itself mediated by a set of technical interventions and interpretations of how they have been disclosed. The content of any such experience is dependent upon the kind of machine within which it occurs and, as such, it is dependent upon interventions and interpretations made during the historical construction of that kind of machine. For example, an "electromagnetic field" is a techno-phenomenon that is not only dependent upon the existence of electromagnets and "electric current" production machines (these, in turn, are dependent upon metal production techniques and chemistry, and so on) but is also dependent upon the practices within which it obtains its meaning. This involved innovating and utilising specific techniques, representations, and interpretations of how machines such as electromagnets work. The electromagnet is a technological object available to produce an "electromagnetic field" only as the result of considerable efforts by experimentalists such as Oersted, Davy, Faraday, et al. It was disclosed through a long historical labour process involved in producing and integrating interventions and representations together, into a single unified technological object available for use. However, contra Hacking, we cannot base a realism upon this stable instrumentality. The performance of any technological object, as a productive object, is itself dependent upon explanatory accounts of that performance and what it has been taken to produce. Thus "spraying electrons on mobidium spheres" is an act of interpretive reference to a manipulative technique made in relation to a machine built in order to disclose "fractional charge". "Fractional charge" is itself an index for a set of particular machine performances that would achieve their theoretical significance, as instances of "free-quarks", through the embodiment of theoretical significance in the selection of techniques and technological objects collected together to construct the machine in the first place. The teleological positings and anticipations that pre-empted the experiment from the onset shape the interpretations of performances of such machines. It does not follow from the stability of those machine performances that the interpretations of them are correct. That is the very question at stake. Given the interpretive dimension to machine performance, technical accounts, techno-phenomena, and technological objects should not be divorced from one another. Techno-phenomena, such as the "electromagnetic field", are defined in terms of what they do, their functions and interactions in specified contexts, and, as such, the concrete character of their performance is inextricably bound-together with technical interpretations of that performance. The "electromagnetic field" is neither purely abstract nor purely concrete. It is both. The performance and accounts are made "hand in hand" through their concrete implementation in the particularities of material practices. Each techno-phenomena is a set of complex machine performances (voltages, time-signals, frequency resonance, etc.) unified under a single index (i.e. electron, charge, repulsion, energy gap, field change, etc.) in such a way as to link theoretical interpretations with technical interpretations of those machine performances. As Gooding pointed out in reference to Faraday, physicists do not respect a neat distinction between contemplative, theoretical aspects of practice on one hand, and instrumental and material practice on the other.[8]

Of course the scientific realists and positivists will object at this point: Does not lightning produce an electromagnetic field? That is the question. Without wanting to conflate epistemology with ontology, I would like to ask another: How do we know that lightning produces an electromagnetic field? "Physicists have measured it!" reply the scientific realist and the positivist in unison. Therein lies the rub. How have physicists measured "it"? How is scientific research done in practice? For Heidegger, the first essential characteristic of research is the projecting a "ground plan of natural events" over Nature in such a way as to sketch out, in advance, the sphere of research in which procedure can operate and provide knowledge.[9] How is this done? Heidegger did not give any account of this. How are the sphere of research and procedure connected by the advance sketching of a fixed ground plan of natural events? The work of Faraday provides a very good example of the establishment of this connection. Gooding's work shows that the development of the stable communicable results of the work by Faraday *et al.* is a history of the considerable effort involved in the development of stable craft practices and representational techniques. Gooding argued that the experimenters were engaged in a process of developing communicable and stable representations that enabled reasoning and skills by conferring meaning upon actions, materials, instruments, and procedures. He deconstructed the orderly reconstruction of the post-experiment narratives of the nineteenth century physicists, which are presented in publications of results and notebooks, in order to recover the processes involved in generating order in the face of the phenomenal chaos of novelty. His analysis showed that these narratives and representations emerged as a result of non-verbal material practices directed towards the construction of cognitive representations through the refinement of those practices. He argued that the theory of electromagnetism was made, rather than discovered, and it has no fixed, independent, essential Nature that can be accessed independently of the manipulations that are involved in the development of stable practices and representations. He also argued that the phenomena disclosed by Faraday *et al.*, and the effects that these physicists encountered on the way to producing those means of disclosure, are not "mere fictions" either. He proposed a convergence theory of agency, which he termed "asymptotic realism", in which experimental and theoretical practices converge when both types of practice achieve practical success in making models that enable action and communication. He used the analogy with the mathematical asymptote to convey the point that at no time is an independent Nature touched.[10] This convergence is directed towards the innovation of stable, reasoned, material practices and experiences of producing novel phenomena.

Experimentation occurs through planned interventions upon objects in the world that are guided by the experimenter's conception of the object and the world. Through progressive actions, the object, the conception of the object (and the world), and the experimenter's conceptions of how to intervene are transformed. Experimentation dynamically creates new phenomena and conceptions of the projected plan of action. Through experimentation, objects and the dynamic process are brought into being. It was this view of the dynamic, creative, and technically rational process of experimental science that seems to have put Gooding in opposition to scientific realism. Gooding's use of the term "asymptotic realism" to describe the psychologism that results from the achievement of stable processes of refinement seems to belong with the realist notion of "approximation", but what Gooding has done is to highlight the extent that the notion of "approximation" is itself only meaningful within the context of both a history of refinement and also a projected future to an unattainable limit. Experimentation aims at objective knowledge, but what constitutes such knowledge, and rationality itself, must be learnt along the way. A change in experimental practice may involve a change in understanding as to the nature of knowledge and its method of acquisition. Furthermore, any understanding of any measurement can only be developed, through experimentation, in relation to an understanding of the techniques by which that measurement was made. It is for this reason that statements of the degree of precision (and confidence in those statements) are linked to evaluations of the sensitivity and "cognitive value" of the techniques used. This remains perpetually open to future refinement and transformation. In my terms, this reveals the extent that mechanical realism is a projection of the metaphysical anticipation of the becoming of the "bringing-forth" of a perfection that is never achieved in practice. It is in this context that the goal of experimental practices can be taken to be the achievement of its own *techne*. Objective knowledge is an imagined ideal that is associated with *techne* on the horizon of a projected anticipation of the causal explanation of the results of material practices. The idea of experimental physics exploring a reality independent of it is only possible from a removed and abstract level of anticipating the theoretical understanding of the completed process of production. However, the completion of this work remains perpetually deferred in favour of the challenge of developing a deeper understanding of the process of production as experimenters attempt to realise reasoning and manipulative possibilities in future experiments.

Progress in research is a technical, functional, pragmatic, revisable, and creative goal of all scientific activity directed towards a perpetually emergent and idealised anticipation of objectivity. Claims to increased accuracy in measurement are based upon theory-dependent techniques and they cannot be compared to any absolute standard. The pragmatic justification of such claims is based upon the convergence between theoretical and experimental practices in which the measured quantity is involved. The basis for pragmatic judgements of convergence are based upon a conception of rationality in terms of a bounded and evolving technical rationality that was directed towards the ideal achievement of complete causal account of the activities of experimental and theoretical practices. Consequently, the standard by which scientific practitioners judge their own objectivity is in reference to "the cognitive value" of their own judgements within an instrumental context of "making". This objectivity requires a social agreement between all (similarly placed) experimental practitioners and is made through the innovation of novel modes of reflection, discourse, representation, and material practice (and not through immediate intuition). This has nothing in common with the immediate experience required by philosophical empiricism for knowledge. Scientific experience is circumscribed as being that which is disclosed through publicly accepted techniques of manipulation and representation. Objectivity, as a socio-technical pursuit, stands in opposition to the "self-evidence of experience" because it must be demonstrated to another by using a mutually understood technique. Experimental physics, as a historical phenomenon, is itself constantly undergoing change, in its theories, objects, and techniques, and change is an essential part of the rational process of scientific inquiry. Subjective experience may well have a role in instigating change but that change could not become part of science until it had been publicly justified via accepted techniques. Thus an observation made by an experimenter can only become part of science once it had been justified to the experimenter *and others* in terms of repeatable observational techniques.

It seems to have been for this reason that Gooding rejected the positivists' appeal to perception, because what was required for "objectivity" and "rational discourse" was a justification of any perception made in relation to technique. Even at the level of measurement there is always the possibility of future refinement and the development of new techniques and instruments. There is no such thing as "fixed data" because "data" is acquired through the use of techniques and there is no such thing as a technique that cannot be refined. Empirical inductive reasoning requires the applicability of concepts to objects and, thus, if empiricism is to be successful, it requires the successful and complete refinement of those concepts in relation to the objective world. However, on Gooding's account, such a process of refinement is never complete and the empiricist is dependent upon the work of others. His argument against positivism (and classical empiricism) was that they have misunderstood the practicality of theorising, neglected the relevance of knowing-how to knowing-that, and that the interdependence of know-how and know-that is just as necessary to defending empirical claims as it is to explaining their origin. Experimenters are not engaged in simply (and passively) registering what is objectively the case because they are actively using techniques, and making judgements about which technique to use, when performing experiments and making observations. It is for this reason that experimenters write down accounts of which techniques they have used. Scientific journals would not accept a paper that merely recorded "observations" without reference to techniques. Experimental observation requires the development of observational skills and if others are not able to acquire these skills then it is unlikely that the experimental observations will be widely accepted. Consider the case of "cold fusion". Fleischmann and Pons were unable to publicly provide a repeatable technique of how to observe "cold fusion" and, consequently, the validity of their work was brought into question. Collins also made this point about experimental efforts to observe gravitons (and also emotional responses in plants).[11] The experimenters' inability to publicly provide a reliable observational technique undermined the public confidence in their observations. When physicists attempt to experiment upon novel phenomena they need to be able to understand one another and so agree about the object of their investigation. This involves coming to an agreement about what phenomenon is under investigation, what they want to learn about it, and how to proceed to learn that. These decisions are made as the investigation proceeds and are not completely fixed in advance. Experimenters learn how to articulate their experiences of novel phenomena along the way of experimenting upon them. Experimental investigation involves the progressive organization of the research, the techniques, the resources, and the descriptions of experiences.

It is only at the point of the asymptotic (unreachable) point of perfection that the object under investigation could be considered to be completely understood, absolutely stable, and functionally repeatable. However, working scientists do not require this impossible degree of rigor before they consider an object to be understood. They tend to consider an object to be understood when it becomes a stable and robust technological object available for future use. It ceases to be an object for experimental investigation, except as an object used to investigate other objects, and, in Heidegger's terms, it becomes standing-reserve. Even when a novel experimental phenomenon is not theoretically understood it still can be known under publicly available technical descriptions. Both the intentions and the techniques implicated in the experimental set-up (the construction of the experiment, its operation, and its theoretical significance) can be known without a complete theoretical description of the phenomenon. Otherwise there would not be any point in performing the experiment. Nor would the experimenters be able to anticipate the phenomenon and devise a plan of action. For Heidegger and Gooding, experimentation is founded on a projected plan of action, which anticipates the phenomenon because novel phenomena require novel forms of communication and representation in order to reassure the experimenters that they are experimenting upon the same thing. This involves producing agreement about the methods of experimentation and also about what was experienced when those techniques were implemented. As Gooding argued, if we are to grasp what a scientific object is (i.e. an "electromagnetic field"), as an object of knowledge, then we need to know how that object has been cognitively engaged with and how cognition was achieved. At each stage of experimental research, the experimenters publicly tie together techniques (both manipulative and representational) and techno-phenomena that are brought into the public realm through those techniques. This involves a progressively developing refinement of "the ground-plan of Nature" as part of the reiterative process of drawing up a plan of action for how the research is to proceed.

For Heidegger, by projecting the ground plan, procedure is provided with a set of objects (an object-sphere) appropriate to that procedure. An example of an object-sphere would be the phenomena under investigation during an experiment on electromagnets, such as coils, wires, batteries, and magnetic needles, and the deviations in magnetic needles when they are moved adjacent to the connected wires. The procedure would be set of techniques for connecting these objects, investigating the contours of the deviations, and mapping them. Modern physics anticipated and projected the plan required for the procedural knowledge of Nature and defined rigour in terms of its exactitude because it must restrict itself to calculate with precision and remain within the object-sphere. Nature, anticipated by the projected ground plan, became the "self-contained system of motions of units of mass related spatiotemporally."[12] In the Newtonian system, motion was defined as the homogeneous and isotropic change of position in a projected grid of space and time, while force was defined in terms of the magnitude of change of position in this grid. Every event in Nature was defined in advance as an event only in terms of how it could be made visible within this projected ground plan. The projected plan was guaranteed by restricting research to the projected plan in every one of its questioning steps. All events had to be defined as magnitudes of motion and changes of motion within the projected grid of space and time that were quantifiable through measurement and calculation. In this respect the experiments in electromagnetism were a continuation of the template of the Newtonian system. However, what Heidegger failed to appreciate was the fundamental novelty of those experiments. The Newtonian system of dynamics was the product of the Galileo's abstraction and projection of the six simple machines within the Euclidean grid of space and time. The electromagnetic machines of Oersted, Davy, and Faraday were not the mechanical devices familiar to Galileo and Newton. They were a novel machine-kind. Did they require a novel ground plan? Was Heidegger too much in the sway of Newton (and Heisenberg, for that matter) and the view that physics is a mathematical science? After all, Faraday is famous for not being trained in the use of mathematics and for being an exemplar of a modern experimental physicist. Was Faraday an exception? Or does Faraday show that physics is not actually mathematical in the common usage of the term? How did Faraday project his ground plans? In order to answer these questions we need to take a closer look at how Faraday developed representational techniques.

Construals, technographe, exoframing, and mathematical practices

Gooding analysed the processes by which Faraday was able to visualise "invisible" phenomena in terms of *construals*.[13] Gooding's argument was that experimenters intervene in "the natural world" and construe their experiences to *create* the correspondence of representations to experience. Observers with different theoretical predilections can agree about salient aspects of the phenomena whilst disagreeing about their theoretical significance. How? Exchanging tentative visual and verbal constructs about "the observed" and how to make observations negotiates agreement between similarly place observers. Observers publicly construe and re-construe their experiences in relation to the construals of other peoples' experiences. Construals are pre-theoretical, practical, situated, and visually representative means of interpreting novel experience and communicating trial interpretations. For example, when we picture and describe light as "rays", "waves", or "vectors" when showing someone how to observe the polarization of light through filters then we are using construals. They are a tentative and public means of visualising and describing an otherwise "invisible" phenomenon (i.e. the motion and axes of light). Construals permit observers to have common (commensurable) experiences of phenomena. The acts of making novel experiences of novel interventions intelligible, such as the motion of a magnetic needle around an electric wire, need to be ordered, either in "real space and time, by moving a real needle around a real disc" or in an imaginary geometrical space.[14] The visual record, in drawings, sketches, and geometrical diagrams, provided the means by which personal experiences could be construed in a form available to public experience. As a form of making order of real-time processes, construals provided the content of the ground plan projected over the phenomena during the setting-up of further experiments. In the case of the early experiments, this ground plan was not the motion of points of mass upon a space-time grid but, rather, the construed motion of the tips of magnetic needles, iron filings, and electric wires upon a space-time grid. This was the projection of a new machine-kind. It involved the space-time mapping of the interactions between moving a needle around a wire and the movement of that needle in response. It involved mapping-out these novel contours of human interventions and machine performances within the technological framework of the mathematical projection of the six simple machines over a new collection of objects. The construals that were used to map out these contours were circles, tangents, arrows, push-pulls, rectilinear motion, anti-parallel motion, and skew motion of the magnets moved near wires and coils. This projected ground plan was essential for the trans-formation of a magnetic needle into a technological object. It could become a probe (or a sensor) and its construed motions could be re-described in terms of "sets", "tendencies", "pointing", "dipping", etc.

Gooding observed, in reference to Faraday's notebooks, that Faraday was aware of the problem of recollecting how he had construed previ-ous experiences.[15] Faraday devised a tactic to deal with this problem. He invented instructions, techniques, on how to construe his previous construals in such a way as to make those experiences stable and repeatable. For Faraday, construals were the interpretive possibilities of motion. These developed against a background of the regularities that he learned to produce. The construals of motion – as provisional and flexible interpretive possibilities – can be compatible with several theo-ries or with none. They enable the earliest (pre-theoretical) stages in the interpretation of novel phenomena and have a heuristic function as a technique for exploring an emergent phenomenal process. Con-struals are selected on the basis of their heuristic, communicable, and instrumental value. Construing involves a complex process of relating actions and imagination. It links phenomenological and abstract space in which both are distinguished, through the construal, in relation to the other. Construals make motion a techno-phenomenon. Attention to the use of construals in communicating techniques and experiences highlights the pictorial (rather than linguistic) aspect of scientific imag-ination. The judgements regarding how phenomena should be repre-sented are socio-technical judgements (made in relation to both other people and material practices) regarding the intelligibility of any techno-phenomena and how they are produced. Consensus between experimenters, dependent upon the successful exchange of observa-tional and manipulative techniques, involved the dissemination of qualitative and pictorial representations of the phenomenon-as-a-process. As Gooding argued, construals are central to the processes of experimentation and they do not permit either a monolithic fit with articulated theories or a metaphysical commitment to determinism, realism, or materialism. However, this raises an important question that, given the peculiarity of Faraday's lack of mathematical training, Gooding did not address: how do construals link with mathematical practices and models?

Jacque Derrida pronounced that specific styles of writing, or inscrip-tion, were required for science to be possible.[16] However, Derrida neglected mathematical forms of inscription and these are central to the present discussion. I use the term *technographe*, as a modification of Derrida's term *graphe*, to denote any physical marks used in mathematical inscriptions, schematics, or diagrams used *technologically* in the design, interpretation, and operation of machine performances. A drawn circle, Arabic numerals, an equal sign, a differential operator, vector notation, matrix notion, Feynman diagrams, electric circuit diagrams, and co-ordinates, are all examples of inscriptions constructed using technographe. If modern physics is inherently mathematical, as Heidegger argued, then we need to understand how mathematics, cognition, and visualization processes are connected. Technographe are used for writing down mathematical techniques and inscriptions. They are the parts of mathematical writing used in constructing solutions, demonstrating proofs, calibrating mechanisms, modelling the performance of machines, and for designing machines. Geometrical proofs, algebra, analytical differential calculus, vectors, matrices, statistics, etc., are all written down, recorded, printed, and disseminated, through the use of technographe. Technographe are not the mathematical techniques themselves, in the same way that *graphe* do not tell us how to write or how to read. They obtain meaning as part of the inscription and interpretation practices used in those mathematical techniques. They are situated within the technological framework embodied in skilled practices.

Euclid's first proposition, to construct an equilateral triangle by intersecting two circles, in *The Elements* is not a logical proof at all.[17] Formally, in terms of modern logic, Euclidean geometry is incomplete because it lacks a continuity axiom in either the Postulates or the Common Notions. The first proposition remains unproven because it has not been demonstrated that the two circles actually intersect. From the perspective of modern logic, the Euclidean geometry available from antiquity to the nineteenth century was not a complete logical system. It was an art rather than a science. The basic postulates of Euclid's geometry, such as to describe a circle with any centre and distance, draw a straight line from any point to any point, etc., are distinct technographe that can only acquire their meaning through repetitive practice. Euclid's geometry is a form of writing in which a set of primitive inscriptive practices constitutes the basis of the whole corpus. Proposition 1 provided the technographe to inscribe an equilateral triangle. This was used to construct further technographe. For example, Proposition 1 was used as a technique in the construction of Proposition 2 which, in turn, was used in the construction of Proposition

3, and so on. These practices are comprised of the inscription acts involved in the mathematical inscription of geometrical figures. In turn, each geometrical figure, once inscribed, becomes a distinct technographe that is used to inscribe further geometrical figures. The first proposition is inscribed by performing the technographic acts of drawing straight lines and circles. A straight line and a circle are defined by Euclid in terms of acts of drawing and, consequently, we can only learn how to perform these practices by following instructions, performing the inscriptive act, and being informed that the resultant is correct. Each figure is a socially mediated artifact. It is a technological object available for further use, and only achieves its truth within the artifice of Euclidean geometry as a set of tacitly embodied practices and their products. We are only able to intuit the self-evidence of these products once we have acquired the requisite artifice and have become mathematical practitioners. Once this artifice is acquired, through education, then practices, reasoning, and intuitions are ordered within its framework.

Each use of technographe is an exemplar, in Kuhn's sense of the word, because it constitutes a set of problem solving tactics that are learned, or constructed, by using them to solve problems.[18] These exemplars were technographically used in *The Elements* to construct geometrical treatments of angles, straight lines, ratios, circles, curves, areas, and solids. The self-evident correctness of these exemplars is established by being able to use them. Each proposition is proved by the act of inscribing it. Its self-evidence is a resultant of its practice and, consequently, Euclidean geometry is as eternal and universal as the social acceptance of the inscription practices upon which it is based. Each axiom is an abstraction and reification of a set of inscription practices. Its status as an *episteme* is achieved by its acceptance amongst its practitioners (and anyone else that they can convince) on the basis of claims for its completeness. These *epistemoi* are collected together and integrated within a technical system as a fixed technological framework with a specified object-sphere (the geometrical figures, proofs, and theorems) and a clearly defined set of interpretations as to how to combine and relate them. Euclidean geometry is characteristic of a *techne* and an *episteme*, from the platonic usage, and an *episteme*, in Foucault's use of the word, as being a total set of related inscriptive practices that is socially presented as eternally, universally, and necessarily true scientific knowledge. They are discursively and technographically related via their embodiment in practice during education. In this sense, Euclidean geometry is an enduring *techne* that has been discursively presented as an eternal *episteme*. *The Elements* provided the *technographic exemplar* for the works of Archimedes, Apollonius of Perga, Nicomachus of Gerasa, and many others. This can be seen in the geometrical proofs of Archimedes and Apollonius, and Nicomachus' study of arithmetic based upon Euclidean ratios.[19] These works, as well as *The Elements*, were preserved and disseminated from antiquity, through the medieval period, and into the present day. The structures of these geometrical treatments were organised within the Euclidean template of axioms, postulates, propositions, corollaries, theorems, and proofs; they provided the exemplars for all subsequent geometry to emulate. Archimedes, Apollonius and Nicomachus innovated new technographe and extended the Euclidean *techne* to include irrational numbers, projections, powers, series, and the geometry of ellipses, hyperbola, and parabolas. The science of mechanics and the *techneic* use of *technographe* to inscribe the motion of the simple machines developed within the same technological framework. As I argued in this last chapter, the conflation of Euclidean geometry as a *techne* and *episteme* (in the context of the construction of the science of mechanics and the desire for novel technological powers) led to the emergence of mechanical realism, and the possibility of experimental physics.

The axioms of geometry have application in our world of experience because, through artifice, we have mathematically projected that application onto parts of the world. It is through the embodied *praktognosia* of pre-conscious habitual familiarity with the techniques of geometry that we are able to "intuit" the applicability of this projection.[20] These axioms are abstractions within a technological framework that we impose upon parts of the world. This is not a structure of our un-trained consciousness; it is the structure of the imposition itself. It is the structure of the mathematical inscriptive practices of the artifice of geometry. If *techne* resides "in the soul of the craftsman" it has been inscribed upon that "soul" through training, mimicry, and practice. The "soul of the craftsman" is not something that we are born with. It is something that we embody through the social acquisition of a tech-nological framework to the extent that we become so familiar with its practices that we are no longer aware of it. Our educated bodies have become situated within the technological framework of geometrical inscriptive practices. That framework, once embodied, becomes part of us and we become part of it. The art and the artist reside in the same place. Projecting the trained imagination performs the "outer sense" of mathematical projection. Our ability to have *a priori* knowledge of the truth of the axioms of geometry is dependent upon the invisibility of technique itself. Our capacity to ground geometrical imagination in self-reflective knowledge is itself a manifestation of the pre-consciousness of technique to a being that is already well versed in the application of that technique. If the technique is invisible then all we see is the projection. It was for that reason that Kant located the origin of that projection in the structures of pure intuition rather than in the art of geometry itself.[21] He was sufficiently skilled in the art of geometry to the point where the art became invisible and its practice became innate.

The mathematician's demand for formal rigour is always the demand for the formal explication of non-formal practice. It is the demand for the logical abstraction of reified skilled human labour. However, the formal axioms, the logically abstracted and encoded system, will be both unintelligible and useless without the skilled practices from which it was reified. A logical proof for the constancy of the ratio of a circle's circumference to its diameter is meaningless if we are unable to draw and recognise a circle. It will not do to appeal to pure conceptualizations of a circle either because without the skilled practices from which those conceptualizations were abstracted there would be absolutely no possibility of applying them in the world. Application requires practice. The technographic practices of geometry (drawing using a straight edge and a compass) were reified in response to the degree of unreflective application of those practices. This reification was itself the product of the challenge of imposing "rational order" upon practices, which were both informal (lacking in rigour) and disordered (inconsistent). The abstraction of those practices was an simplification of what was already taken for granted as being technically correct, in such a way as to induce a universal *techne* under which those practices would be integrated under a single theory of practice, which paid no attention to the particularities of practice. The move towards abstraction is an attempt to detach the technique from its application in order to generalise it from the particularities of use. This involves a synthesis of a diverse and divergent set of particularities into a general axiomatic system that can be applied equally to all those particularities. Under the *techne* of geometry the objects produced by the application of that *techne* would be abstract, idealised reconstructions of technographe that were conceptualised in terms of definitions generalised as axiomatic principles. This reification allowed the axioms of geometry to be divorced from their practical origins and contexts of use. However, this required the techniques (and practices) of logical abstraction into general rules which, in turn, also requires abstraction if the formal system is to be a complete *techne*. All mathematical abstractions, as encoded reifications of techniques, are incomplete. This is the source of all error and problems for the application of a formal system to experience. It is not the case that the formal system does not correspond to reality (as the realist would argue) and hence the error. It is more the case that the system is itself incomplete and we are not in possession of a full and complete *techne*. In short, we do not know what it is that we have done, nor do we fully know how to apply the general system to particular experiences. We are forced to keep experimenting or give up. Furthermore, without completeness we are unable to unify the whole abstraction into a formal *techne* and, consequently, we cannot consider the particular practices from which the system is abstracted to be homogeneous. They are a heterogeneous collection of exemplars and tactics. As I shall argue in chapter six, it is this heterogeneity that is the source of all "error" and "resistance". This is a matter of coherence rather than correspondence. The challenge of system building is to collect together and integrate heterogeneous objects into a homogeneous whole. The inconsistencies that arise during that process are the results of the interference between heterogeneous objects synthetically brought together in the attempt to integrate them into a novel unified and coherent composite. The "failure" of any system is a consequence of its incompleteness, complexity, and disunity.

The invention of non-Euclidean geometries brought with it a novel awareness. Not only were geometrical objects the products of our mathematical practices, but also they were arbitrary. There is nothing unique about any particular set of technographic practices; dimensionality can be projected in any number of different ways. Any numbers of new technographic practices were possible (as Boltzmann, Lorentz, Minkowski, Gauss, and Einstein have shown). We are free to construct any topology from any arbitrary set of axioms. That topology is considered rigorous, in the disciplinarian sense of the word, providing that our axioms and definitions do not contradict one another when we attempt to combine them. There is not any "objective" space to which mathematical geometries must "correspond"; topology can only provide us with a coherent system of consistently mapping arbitrarily imposed axioms and definitions. The problem is how to relate these arbitrary spaces to the practices of experimentation and measurement. Once again, this raises the question: How is the "empirical" practical dimension of experimentation connected with the "rational" mathematical dimension? The case of Faraday's work provides an illuminating answer to this question. If we accept Gooding's argument that the use of construals was central to Faraday's public reasoning process then we can readily see that this process was a technographic process. Socio-technical judgements are premised upon technographic cognitions and it is for this reason that I agree with both Gooding and Heidegger. Such cognitions are made within the context of a socio-technical learning process and realised "for ourselves" as something brought by us to the learning experience. This is not an innate cognition but is a socio-technical cognition that is made "ours" through the embodiment of artifice through innovative practices. It is premised upon an open and praktognosic orientation towards objects from within the invisible technological framework of embodied artifice. Thus Faraday's work was not mathematical in the common sense of using mathematics, but it was mathematical in Heidegger's sense of *mathesis*. As Gooding pointed out, "construing creates 'giveness' of experience... [It is] a relatively stable but plastic interpretation of experience which guides further exploration and interpretation."[22] It is this "giveness" that is characteristic of *mathesis* and mathematical projection. It is the "laying down of the ground-plan of Nature" in retrospect whilst leaving it open for future refinement.

Ampere's key experiments, as Gooding noted, were designed to preventing movement and reduce complex interactions towards stable equilibrium.[23] Why did Ampere choose this particular configuration? There are two possible reasons that I would like to discuss. The first possibility was that he had adopted a mathematical tactic, in the common sense. This tactic was set-up to avoid the practically enormous task of finding solutions to the mathematical expressions for the complex phenomena produced by the "process-structuring" techniques of Oersted, Davy, and Faraday. In my view, this tactic was the attempted extension of the Galilean reductive method of reducing all mechanical motion to that of the lever and the balance. This would have allowed Ampere to treat the problem as if it were one of ratios and simple differentials. The second possibility was that Ampere had adopted a mathematical tactic, in Heidegger's sense. He projected the Galilean template. Faraday, on the other hand, was destined to non-mathematical work, in the common sense. This would put him in the "empirical tradition", according to Kuhn and Osler. And yet, in Heidegger's sense, Faraday's work was an exemplar of mathematical projection. How did Faraday know how to begin? According to Gooding, Faraday's plans of action began in October 1820 with attempting to reproduce and map out the contours of Oersted's famous needle-wire motion observation.

After eleven months of experimenting with sideways motion, circular motion, and push-pulls, he managed to stabilise his configuration of the 1821 compact rotation apparatus. This produced revolutions! It was the stable predecessor of the electric motor. It was a hybrid electrical and magnetic machine that produced stable rotations when it was connected to a chemical battery by metal wires. It both enabled and constrained the spontaneous circular motion of a needle pendulum around a magnet. What constrained that motion? As Gooding observed, the motion of a suspended magnetic needle around an electric wire is far from stable and well defined. It takes considerable patience, practice, and skill to manoeuvre a needle around the wire (without it touching the wire) and map out its motions. One must control one's interventions carefully. How did Faraday plan the control of his interventions? He did so by exploring the possible motions of sideways motion, circular motion, and push-pulls, and with meticulous attention to those possibilities, Faraday mapped out the contours. The 1821 machine was designed to demonstrate circular motion and an otherwise quite unpredictable movement was constrained to a circular path. It is in this sense that Faraday mathematically projected the abstracted circular motion of the wheel and axle as the ground plan of the possibly stable motion to try. Faraday's experiments on the motion of magnetic needles and electric wires were constructed to capture the circularity of that motion. By doing so, Faraday projected the ground plan of circular motion, but this did not adequately capture the movement of the "rotations". As Gooding pointed out, Faraday's later construal of these "rotations" as skewed motion were not a Newtonian construal of force, and "Faraday took up the possibility that the skew-aspect was not to be reduced."[24] He deconstructed (via Ampere) the applicability of Galilean reduction to the novel electromagnetic machines. A new construal was required. For an example of Faraday's construal of "skew" motion see Thomas Martin's transcription of the first part of Faraday's experimental record for 3rd September 1821.[25] If we examine this skewed motion then it is clearly evident that it is an abstraction of the screw. Faraday managed to resolve his difficulties in construing that motion by using the screw as his construal. The screw was a non-reductive exemplar of "electromagnetic motion" and it was still one of the simple machines. It also allowed mathematical techniques (in the common sense) to be imported from the six simple machines to the novel kind of machine. Maxwell's field theory utilised Faraday's construal of the tangential, or skew, motions of "electromagnetic lines of force" in terms of the screw. Faraday's construal of his experiences in terms of screw motion of the "invisible" lines of force is a breaking free from the Galilean reduction. The mechanical circular and rectilinear motion would not suffice for the novel machines. Faraday's screw-construal was a non-reducible primitive. Once Maxwell had invented the *grad* and *curl* operators of differential calculus (and specifically invented for this task) then such motion could be described in terms of differential calculus. How? So far, I have only discussed the pictorial and geometrical technographic inscription of machine performances. One of the crucial mathematical innovations of the seventeenth century was analytical geometry and differential calculus. These allow machines to be analytically inscribed. How are technographic inscriptive practices connected with models, interventions, and representations?

Functive was a term used by Gilles Deleuze and Félix Guattari in their analysis of science and its distinction from art and philosophy.[26] A functive is an element of a function. Limits and variables are examples of functives. A function defines the relationship between its functives and may, in turn, be used as a functive within another function. In physics, a function has to refer to a co-ordinate system in which the axes represent physical quantities. In Deleuze's terms, physics proceeds in the face of the infinite chaos of existence by constructing a plane of reference from co-ordinate systems in order to slow down the disorder of chaos by external reference (or *exoreference*). Exoreference involves extrinsic determination of the meaning of the frame of reference. Physics is distinct from mathematics due to this extrinsic determination. Mathematics finds its meaning in the intrinsic consistency (endoreference) between functives, whereas physics has to extrinsically give those functives physical meaning. Exoreference allows functives to participate in modelling and provides a co-ordinate system, composed of a least two independent functives whose relationship is the function, with meaning as a state of affairs or informed matter. Exoreference is necessary for the frame of reference to form a proposition that relates a state of affairs meaningfully to the system in question. For example, it is an act of exoreference that is required for the differential functive dy/dx to refer to the rate of change of pressure with respect to temperature. This gives it meaning as a state of affairs between pressure and temperature in a system that is extrinsically determined as a sphere of gas at constant volume. The function allows each dimension (axis, variable) to be fixed whilst the others are varied. Exoreference allows mathematics to participate in modelling machine performances.

Functives and technographe are meaningless without the technological framework of inscriptive practices and exoreferences in which they are situated. Mathematical methods (as procedural collection of mathematical techniques) utilise techniques and ordering technologies as part of that technological framework. This framework is also associated with its application. For example, Fourier analysis involves a collection of different technographe, functions, functives, techniques, and inscriptive practices. It is technique for analysing complex "wave patterns" in terms of series of sine and cosine functions. In order to be effective it must be embodied in inscriptive practices and exoreferences in the context of solving a range of particular problems. By applying Fourier analysis to the solution of an inscribed physical problem, say the solution to the Schroedinger Equation for electrons within a metal wire under a potential difference V, the solutions of this technique can be exoframed as expressions of physical states. The sine and cosine functions of the Fourier series can be taken to be the wavefunctions that are superimposed to probabilistically describe the measurable behaviour of the electrons. By inscribing the contours of the interactions between human interventions and machine performances, in terms of functives by using technographe, the contours can be mapped out in terms of operational and responsive variations. Exoreferences give this mapping meaning. Hence, the physicist can slowly turn up the pressure acting on an experimental cell and read the variations in temperature from a calibrated thermometer. This can be recorded graphically and written down in the form of a differential equation. Differentials are particularly suited for inscribing machines and mechanisms. The differential relates variables in terms of a ratio of a rate of change, of one with respect to another. These variables can be used as dimensions to construct an analytical external framework in which the differential equations can provide the contours of the physical process under investigation with both qualitative and quantitative meaning. It does this in terms of exoreferences that are projected over the machine performativity in terms of fundamental variations and dimensions. As a part of a technological framework, these projected exoreferences are associated with the operational practices of the experimental procedures, and hence the projection of this framework is an anticipatory act of *exoframing*. The process of exoframing facilitates the process of writing out, deinscribing, both human interventions and machine performativity from the final products of experimental work. The technographic inscriptions can then be taken to be representations of the physical processes involved in the experiment. Both human interventions and machine performances are de-inscribed (written out) from the final accounts. The book of Nature can then be read.

"Making" the ground plan of nature

The work of Faraday was an example of exoframing the novel phenomena and modification of the mathematical projection to allow more freedom to construe the screw motion without the restriction of the Galilean reduction. Faraday *et al.* had invented a novel kind of machines and confirmed the methodology of mathematical projection. His work is an example of how the so-called "rationalist" and "empiricist" traditions were both aspects of the same technological framework. He projected his ground plan of action in order to establish a series of interventions by which he could map out the phenomena by building machines. This provided Faraday with a general methodology by which he could set in place his object-sphere (his magnets, wires, needles, etc.), his set of possible construals (rotations, screws, antiparallels, etc.), and his procedure (build a machine and map out its motions). The rules and laws could then be abstracted in terms of the technographic maps of the motion of the machine (the "lines of force" of the "electromagnetic field"). According to Heidegger, procedure must set in place the changeable and allow it to change, as a sequence of calculable events, in order to allow facts to become objective, fixed, and, hence, determinations of the constant and reproducible aspects of the changing of the changeable can be made into rules.[27] Again Faraday did this. Faraday construed the movements of magnets and needles as the changeable motions of the six simple machines by constructing the apparatus in such a way as to prevent it from making any movement that was not a simple machine motion. It is important to note that I am not claiming that Faraday was a stage magician who built a trick. I am claiming that he reduced the possible movements to that of one of the six simple machines and that his procedure, building a machine according to the methodology of the template, set in place and restricted the changeable movement to one of the six projected mechanical motions, in such a way as to make it repeatable within the projected plan. This allowed the technographic construals and representations of that motion to be used map out the contours of that motion in terms of constants and variables. These can be used to construct an exoframe and, eventually, a mathematical law. After all the work of exoframing has been done, the "empiricist" will then claim that the law is a good fit to the phenomenon, and the "rationalist" will then claim that the phenomenon is a necessary consequence of that law.

In the context of experimental physics, the changing of the changeable is the stable and reproducible changes of those variable aspects of the experimental apparatus that are under investigation. It is the contours of the interaction between human interventions and machine performances. The process of research is then one of mapping out these contours by using those techniques and technological objects as transformational agents. It is the task of the experimenter to determine, in relation to the contours, which aspects of machine performance change in response to particular human interventions. The process of research is subsequently one of mapping out the changes of the changeable in relation to exoframed human interventions. For example, we could map out the change in volume of a gas in relation to changes in pressure and temperature. This would involve collecting together a piston, a Bunsen burner, a thermometer, and a pressure gauge, and varying each of the variables (temperature, pressure, volume) in relation to one of the others (whilst the third is fixed) to map out the contours of human interventions and machine performances. Those contours could then be presented as the manifestation (resultant, consequence) of the realization of the rules by which those variables are related "in Nature" via a mechanism. Methodology sets down the set of techniques and technological objects that will be used by the research to disclose the "natural mechanisms" to be disclosed by that experimental research. By exoframing those contours, we could then write down the differential equation for those rules. What we have done (or so the mechanical realist would claim) is to create an artificial space (free of the chaos of competing mechanisms) by which those rules could be disclosed. Furthermore, or so the mechanical realist would claim, the differential equations that we have written down are, in fact, a representation of "the laws of Nature" that govern such a process. By fixing these rules as the necessary consequences of "natural laws", methodology is able to determine "the laws of Nature" in terms of the rules governing the changing of the changeable that has been set in place by exoframing the experimental procedures. By projecting the ground plan of Nature, physics has made this so. Facts can only be made clear, as the facts, within the purview of concepts, rules, and laws, and, therefore, research into the facts is intrinsically the establishing, verifying or falsifying rules and laws. In my terms, physics is thus able to perform the "sleight of hand" that has been premised by the pursuit since its origins. Through methodology it engages in a process of making that is directed towards the techneic realization of unchanging principles of change. It has presumed, on the basis of mechanical realism, that these principles are epistemic principles, whilst simultaneously de-inscribing the participation of human interventions and machine performances from that process. It is then able to remove its own methodology (using machines, etc.) from the final account and present its *techne* as "Natural Law".

For Heidegger, research is made objective by encountering it in "the complete diversity of its levels and interweavings", whilst procedure is freely directed to view the changeable in its object-sphere that provides "the horizon of incessant otherness of change" required for facts, as concrete particularities, to become present.[28] This is crucial because human beings do not control the outcome of experimentation. The disclosed would not have been disclosed without human intervention but the disclosed is not controlled by that intervention. What is controlled is the template of the experiment, how it is performed, and this template constrains (but does not control) what can be learnt. This provides methodology with an explanatory character because it can present itself as a mode of disclosure. Methodology accounts for an unknown by means of a known and, at the same time, verifies that known by means of that unknown. Both the known and unknown are used to explain each other. Heidegger did not provide an example of this. A good example is the case of gravitation. Newton's Law of Universal Gravitation explained already established facts (objects fall to the ground when dropped, the Moon orbits the Earth, and the planets orbit the Sun in elliptical paths) in terms of an unknown (the unitary force of gravitation) via a procedure (the calculation of paths using the inverse square law). Hence the unknown could be accounted for in terms of the known facts and these facts were verified as necessary consequences of the unknown. This can also be seen in Faraday's experiments. The known possible motions of electromagnetic machines would be explained in terms of "an unknown force" via a procedure of mapping out "the lines of force". It is for this reason that I agree with Heidegger's claim that "natural law" is established with reference to the ground plan of the object-sphere that provides criteria and constraints upon the anticipatory representation of the conditions under which the experiment can be performed. This set up is required to prevent the representations necessary for experiments from being based upon "random imaginings" and is based upon the ground plan projected onto Nature and the representations are sketched into that ground plan. The planning and execution of experimentation, as a methodology, is supported and guided on the basis of the fundamental law that allows the facts, that either verify or falsify the law, to be induced into a general *techneic* form. Due to the operational precepts of mechanical realism, this *techne* was conceivable as "Natural Law", and the phenomenological object of machine performance could be presented as the means to disclose that law. In modern physics, the ways in which experimentation is performed is dependent upon the particularities of what is being investigated and which type of explanation is required. However, it is only through the transformation in general conceptions of knowledge and Nature that research through experiments became possible. What was this transformation? Firstly, mathematical projection, according to Heidegger, begins with laying down a law as a basis.[29] An experiment does not begin with a complete physical law, which it then tests (as if often presumed) but it does presuppose the existence of a law, which it aims to disclose. Thus, the presumption of the existence of a law is the basis for an experiment. As Gooding argued, the "discovery" of this law, in the form of an empirical regularity or a mathematical function, requires considerable effort on the part of experimenters and may take decades of work before it has been formulated. However, the methodology by which experimental physics operates involves the presumption that there is a law, which can be disclosed by the proposed experiment (or series of experiments), and presupposes the forms of possible motions that would qualify as regular motions. For Heidegger, to set up an experiment requires representations and conceptions of the conditions under which these specific series of motions can be made susceptible to being "controlled in advance by calculation" and followed in their "necessary progression". Heidegger was directing us to the way that physics has pre-empted what could be possibly learnt, to such an extent that it has pre-empted the conceptual and representational conditions under which an empirical regularity would be produced, recognised, and accepted as such. These representational and conceptual conditions also pre-empt how the experiment could be built in such a way as to be a controlled experiment. In other words, it has pre-empted what qualifies as a constant conjunction and the conditions under which an experimentalist could claim to have produced constant conjunctions. It is because a constant conjunction is a repeatable conjunction of events, and each event is construed, in advance, as one of the projected six simple mechanical motions, then performing an experiment inherently and necessarily involves constructing machine performance. The experiment is controlled to the extent that some kind of machine performance will be its consequent but which combination of simples will be disclosed is not controlled. This allows physics to participate in discovery. It discovers the particularities of the machine performance of the machines it has built according to the preconception of how a machine should perform.

Technique binds together an object with procedure and methodology; hence the technological framework of techniques and objects upon which it projects its template defines the methodology of any experiment. Distinct experiments are defined in terms of related sets of technological objects, procedures, and theories in accordance with the teleologically posited purpose of the methodology. Each machine as an associated cluster of techniques, technographe, exoframes, and technological objects, is a distinct object-sphere. The exploration of that object-sphere, by mapping out the possible interactions between human interventions and machine performances, is an experimental research project. This is as true for rats in biological experiments as it is for the motion of needles around wires. Rats are disclosed by technique to be a set of machine performances (repeatable responses to interventions) just as much as the needle and wire were when Faraday assembled them into a rotation device. The methodology through which individual object-spheres are "conquered" does not simply amass results, but uses those results to adapt itself to a new procedure. Experimental sciences are verified by the successful development of its projected plan, by means of its methodology, into new procedures and experiments. As Heidegger pointed out, the specialization of each science into these specific fields is not a "necessary evil", due to the increasing enormity of the results of research, but is a necessary consequence of the practical test of any methodology being defined by its capacity to facilitate the "ongoing activity" of research.[30] Any attempts to characterise science solely as "serene erudition" could not sustain or explain a notion of modern science as an ongoing activity, its performativity, its capacity for enduring, nor the "self-evidence" of its results. Ongoing activity provides any research project with the capacity for institutionalization in terms of its self-appointed task to restrict itself to a particular field of investigation, whilst sustaining its "the solidarity and unity" as a specialization of a particular science. The ongoing activity of research builds the plan of an object-sphere into all adjustments that facilitate any planned conjoining of types of methodology, that further the reciprocal checking and communication of results, and that regulate the exchange of talents and skills. Extending and consolidating the institutional character of the sciences, as an ongoing activity, secures the precedence of methodology over Nature and determines, at any given time, what is taken to be objective in research. It is the adaptation of research to its own results that, as an

ongoing activity, provides modern sciences with advancing methodologies and an intrinsic basis for the necessity of the institutionalization of research. According to Heidegger, institutional specialization
provides the basis for solidarity of procedure and attitude, with respect
to the objectification of Nature that constitutes "the real system of
science."[31] The researcher is directed according to institutionalised and
legitimated projects appropriate to the object-sphere in question. The
negotiations at meetings, the information collected at conferences,
the books and papers contracted by publishers, are all directed and
organised through the institutionalization of specializations. The
research worker is, consequently, forced to operate within "the sphere
characteristic of the technologist" in order to be "capable of acting
effectively". Scientific progress, defined in terms of ongoing activity,
specialization, and predictive success is a matter of the increased technological enframing of phenomena into the framework of technique
and bounded technical rationality. Thus empirical investigation is
itself a manifestation of *Ge-stell* in which each progressive refinement
of accuracy is a standing-reserve for the future challenging of the technological framework to become increasingly precise and exact. The
objects and measurements of "empirical research" are bound-up with
the destining of technique that cannot ultimately be satisfied because
it has no end apart from itself as a means. It is itself an experiment into
its own possibilities as modern physics simultaneously establishes and
differentiates itself in its projections of specific object-spheres. The
development and specialization occurs by means of a corresponding
methodological projections of ground-plans of research that are made
secure via the rigorous application of procedure, which is adapted and
established, at any given time, in ongoing activity. Projection and
rigor, methodology and ongoing activity, mutually requiring one
another, constitute the essence of modern physics and transform it
into research. The unity of this system is not contrived by relating
object-spheres according to their content, but calls Nature to account
insofar as Nature "lets itself be put at the disposal of representation"
and, consequently, is calculated in advance and "set in place" by
research.[32]

The theory of the real

Science is given cultural value and is pursued by human beings from a
variety of motivations. The sciences have intersected with industry,
commerce, education, politics, warfare, journalism, and the arts.

However, asserted Heidegger, we cannot understand the essence of science and its scope if we take science as cultural only this sense.[33] Western European science determines the fundamental characteristics of the reality within which "man of today moves and attempts to maintain himself" and has developed unprecedented power that is "ultimately to be spread over the entire globe". Heidegger raised the question as to whether science is "nothing but a fabrication of man", or whether "something other than mere wanting to know on the part of man", rules in science. Heidegger's intention, from the onset, was to reveal what this "something other" could be. He started this inquiry by equating this "something other" with "a state of affairs" that reigns over all the sciences but is hidden from them. Heidegger proposed that science is *the theory of the real*. What did Heidegger mean by this single concise statement? Heidegger elucidated "the theory of the real" by means of an etymological analysis. He analysed the modern conception of "the real" in terms of "that which works" to show how performing and executing are central to the setting-forth and self-exhibition of reality. Heidegger used *arbeiten* and its compounds (*bearbeiten*, "to work over or refine", *zuarbeiten*, "to work toward", and *unmarbeiten*, "to work around or recast") juxtaposed with *wirken* ("to work"), in order to set in place the performative way in which modern science brings "the real" (as an object in a causal sequence) into presence. Modern experimental science involves working towards and striving after reality in order to capture and secure it. Theory as *Betrachtung* meant capturing, entrapping, and secure refining of "the real".[34] "The real" is ordered into place as "an interacting network" of a series of related causes and, thus, is made into something that is capable of being followed out in its sequences. This secures "the real" in its objectness and provides object-spheres and object-areas for scientific observation and procedure to capture. The work of modern science is one of performing representational captures that refine "the real" in accordance with the objectness through which everything real is recast *in advance* into a diversity of objects available for representational captures. It was this conception that allowed the factual to follow from "deeds and doings" whilst retaining the connotation of certainty. It was for this reason that Heidegger considered that the transformation of the conception of "the real" into "the certain" was characteristic of the post sixteenth century modern age in the way that "the real" has been represented as present in the occurrence of consequences. It is this characteristic of "the real" which provided that kind of presencing characteristic of the modern age that Heidegger termed as *objectness*.[35] For Heidegger, how the objectness of that which presences was brought to appearance, and how that which presences became an object for representing, could only be understood in relation to theory.

How is Nature, supposedly the object of modern science, presencing itself? What is the "it" here? How did Heidegger characterise the fundamental conceptual change that allowed that conceiving of Nature in terms of its objectness? For Heidegger, every new phenomenon emerging within an object-area of science is refined to such a point that it fits within the normative coherence of theory. This view has parallels with Kuhn and Feyerabend. Heidegger maintained that normative coherence is itself changed from time to time whilst objectness remains unchanged in its fundamental characteristics. This idea of changes in normative coherence has parallels with the notion of "paradigm shift" (as coined by Kuhn).[36] How does objectness remain unchanged? What are its fundamental characteristics in modern experimental physics? For Heidegger, if representing in advance is the basis for strategy and procedure then science is determined by pure theory directed towards the objectness of what presences. Thus the validity of classical physics was limited but not contradicted by modern quantum physics and relativity. Modern subatomic and space-time theory refined the object-spheres of their respective researches but did not invalidate them. The narrowing of the realm of validity was a confirmation of the objectness normative for the theory of Nature. How is "the objectness of what presences" chosen in order to set-forth the objectness of "the real" in such a way that there is not any fundamental change in objectness between classical and relativistic quantum physics? In the experimental sciences "the real" is "what presences as self-exhibiting", refining it corresponds to a fundamental characteristic of "the real" itself, but its presence is brought forth to stand in objectness. For Heidegger, modern science, as the theory of the real, was not self-evident, nor was it merely a human construct. Wrought by work, defined by the setting-forth of presencing into objectness, scientific theory challenges and sets-upon Nature in order to disclose "the real".[37] According to Heidegger, it is Nature that "presents itself for representation as a spatio-temporal coherence of motion calculable in some way or other in advance" in accordance with theory. Nature, according to Heidegger, is clearly an amenable participant in the work of modern science. What is the Nature that presents itself in this way? Or, to be more precise, which parts of the world are taken to be instances of Nature presenting itself in this way? And, how is the temporal coherence of motion calculable in advance "in some way or other" for these parts? Heidegger maintained that modern science, by defining "the real" to be the measurable, provides a method that permits a decision regarding what may pass as science by limiting certainty and knowledge to the measurability supplied by the objectness of Nature and the possibilities inherent in the measuring procedure. What allowed "the real" to be defined as the measurable? How does Nature supply measurability? How are the possibilities of measurement inherent in the measuring procedure? Heidegger argued that the methodology of science sets up Nature as an object of expectation. All objectification of "the real" secures and guarantees some coherence of sequence and order. Mathematics participates in this methodology by setting up, as the goal of expectations, the harmonising of all relations of order and "reckons" in advance with one fundamental equation for all possible order, and is not merely a reckoning by performing operations with numbers for the purpose of establishing quantifiable results.

According to Heidegger, the "inconspicuous state of affairs" which conceals the essence of science can be revealed by taking particular sciences as examples and attending specifically to whatever is the case regarding the ordering, in any given instance, of the objectness belonging to the object-area of those sciences.[38] For Heidegger, physics (in which he included macrophysics, atomic physics, astrophysics, and chemistry) observes Nature insofar as Nature exhibits itself as the objectness of coherence of motion of material bodies. The elementary objects of classical mechanics were the motions of geometrical points, in nineteenth century physics these objects were fields and atoms, and in the twentieth century the interaction of elementary particles is the manifestation of the "impenetrability of the corporeal". However, Heidegger, by generalising from Galileo's *Assayer*, Newton's *Principia*, and Heisenberg's positivistic interpretation of quantum mechanics, had not attended to modern experimental physics closely enough, and, consequently, was compelled to consider the source of novelty and the essence of modern physics to be enigmatic. For Heidegger, the essence of science was "rendered necessary" the moment that this setting-upon occurred, but that moment, and its possibility, remained mysterious.[39] However, once we take mechanical realism into account then we can reveal the possibility of this "moment" and its subsequent "necessity" within the cultural demands for power and certainty. If we attend to how experimental physics is actually done, and what it is actually done to, then we can attend to the specific ordering of the objectness belonging to particular object-areas of physics without preconceiving, from the onset, the nature of the object of that pursuit. This object-area is comprised of strata of machines, techniques, and their associated techno-phenomena. The techneic laws abstracted from the alethic modalities of the contours of the interactions between human interventions and machine performances are the determinations of the estimated possibilities, actualities, necessities, impossibilities, and contingencies of exoframed labour processes.

Heidegger was correct to locate epochs of change, such as the change from classical physics to quantum physics, in the experience and determination of the objectness of the appropriate object-sphere, whilst emphasising that the essence of modern physics remains unchanged. However, for Heidegger, "in the most recent phase of atomic physics" the object vanished. Heidegger alluded to a change in the objectiveness of Nature into the constancy determined from out of *Ge-stell* and made reference to *The Question Concerning Technology*.[40] Heidegger noted, in reference to the Wilson cloud chamber, the Geiger counter, and the balloon flights to detect mesons, that modern subatomic physics, despite its aim to make elementary particles exhibit themselves for sensory perception, can only provide indirect, via a multiplicity of technical intermediaries, self-exhibiting of elementary particles. It is this indirectness that Heidegger alluded to as being indicative of the dominance of *Ge-stell*, as a fundamental change in the experience and determination of objectness, in the most recent phase of physics. However, by presuming that the object of experimentation in pre-quantum physics was present to perception, Heidegger revealed his bi-partite structure of experimental physics in terms of law-object. Given that, for Heidegger, law is something that we project over the phenomena in terms of cause-effect relations in order to map out the changing of the changeable, then the objects of investigation remains those aspects of Nature that are amenable to this project. It is an implicit consequence of Heidegger's analysis that the progressive criterion for selection of law could only be one of empirical adequacy of the law's description of the cause-effect relations to the changing of the changeable within the object-sphere. Thus Heidegger marked out "the most recent phase of atomic physics" as something distinctly given over to *Ge-stell* because the objects of its object-sphere are unavailable to direct perception. Heidegger's interpretation of the aim of experimental physics as being directed towards the mathematical projection of laws describing the coherence of motion of its objects, and the constancy of the changing of the changeable, betrays his positivistic conception of science in general and physics in particular. In essence, the aim of science, on Heidegger's account, is that of empirical description in terms of universal law. This should be unsurprising, given that Heidegger's two exemplars of scientific endeavour are Newton and Heisenberg, and that in Germany, at the time of Heidegger's writing, positivism was the dominant conception of science. I am not suggesting that Heidegger was a positivist (far from it!) but his conception of the goal of science was positivistic. Thus, for Heidegger, the object disappears in subatomic physics because it cannot be directly revealed to sensory experience. Here we can readily see how Heidegger's own phenomenological pre-occupation with "that which presences" obscured his view of the phenomena of experimental physics. However, as I have argued above, the aim of modern experimental physics has always been the disclosure of mechanisms, and its ontology has had a tri-partite structure (object, law, and mechanism) since the sixteenth century. Particle physics is unconcerned with the detection of "particles", mesons for example, except as a means of investigating its models of elementary particle interactions. Physicists do not seek the "truth/top quark" for its own sake but, rather, for the disclosure of fundamental mechanisms. Mechanisms have never been brought into presence, as objects, but are central to causal explanations for "that which presences". Since Galileo, physics has had pretensions towards a "deeper" ontological relation with the mapping of temporal successions of cause-effect sequences than merely confirming or refuting the law. These explanations are given as the underlying workings of that which allows "that which presences" to presence. "The real" disclosed by experimentation, via demonstrations given in terms of the temporal sequences of changes of the changeable, is brought into discourse as a causal account given in terms of the mechanisms that are proposed explain the phenomenal. The objectness of "that which presences" is taken to disclose "the real" via the objectness of the object-sphere and the constancy of the changing of changeable as means to the disclosure of "the real" as a mechanism of change. The object-sphere, procedure, methodology, and mathematical projection are tied together via an exoframed model of the mechanisms in operation in the changes of the changeable. It is the projected exoframed model that allows mathematical projection, the object-sphere, methodology, procedure, law, and the ongoing activity of working towards securing representations, to fit together as a coherent process of research. Working novel experimental physics is not limited to sensory experience but, rather, explores what is disclosed by means of sensory experience. Its object-spheres, object-areas, procedures, laws, advancing methodologies, and mathematical projections, have always been means to this end, and, in this respect, it changes the changeable in the object-sphere to suggest causal mechanisms which are "tested" by implementing them in the ongoing technological procedures of experimentation. Thus the use of objects as standing-reserve available for future use and ordering has been central to the processes of "testing" in experimental physics since its onset. *Ge-stell* has been characteristic of modern experimental physics since its origins in the science of mechanics. *Ge-stell* has been operational within the unfolding and ordering of the ontology of experimental physics in all of its object-areas (i.e. mechanics, optics, thermodynamics, electromagnetism, acoustics, solid state physics, atomic physics, and subatomic physics) and is not restricted to the "most recent phase in atomic physics". Heidegger's positivistic conception of modern science concealed the operation of *Ge-stell* in experimental physics and, consequently, concealed the technological essence of modern experimental physics.

Heidegger was correct to have considered objectness to be essential to the setting up of modern sciences in general and modern physics in particular, as a methodologically and procedurally secured mathematical research projection of the ground plan of its object-sphere. Modern experimental physics could have not begun nor operate without this set-up. What Heidegger misunderstood is the objective of the pursuit of experimental physics. It is *techne*, as an ideal, that provides an asymptotic link between practice, theory, and scientific rationality. Knowledge becomes objective upon the creative establishment of the possibility of achieving a *techne* of how that knowledge was produced. Thus the process of scientific progress is one of aspirations towards technical excellence that is to be achieved via a techneic process of questioning and correcting both the content of theories, experimental and theoretical practices, and standards of justification. Gooding termed this process to be one of *convergence*. This open-ended process is one that is perpetually directed towards the future and, as such, is one for which there are not any rules to guide it because we cannot foresee the course of innovative development. It is one of bounded and evolving technical rationality. Controversies regarding the artificiality of the results of techniques and preparations (what is a property of the object and what is a product of the preparation process) could only be achieved by reaching theoretical agreement as a consequence of further technical exploration. This requires both theoretical accounts of the techniques utilised and of the object under investigation. The establishment of a trajectory of investigation is the technically rational conclusion of any controversy. Prior to publication, any experimenter needs to reflect upon the experiment (its purpose and execution) and anticipate any criticisms to her/his work. Every experiment is situated within the background of a wider scientific community's standards and expectations in which it will achieve its significance and meaning. The working scientist will base her/his conception of what makes a reliable observation and/or technique upon recognition of the level and content of any possible criticisms that her/his work may be subjected to. It is this ability of the experimenter to anticipate how others will receive her/his actions and reasoning, which is a necessary condition for the critical self-reflection to be presented as "objectivity", whilst, in proposing any novel theory or experimental technique, it is necessary for the experimenters to break the accepted norms and standards of the current scientific community by engaging in a critically motivated reflection upon the correctness (or legitimacy) of those accepted norms and standards. This requires that the framework that the experimenters are working within is open and capable of being developed and changed. If the framework were fixed, final, and closed there would not be any potential for novel research because there would not be any possibility of establishing new techniques, theories, and objects for investigation. Experimental work is an unending process of producing a series of refinements for which successive refinements correct and reveal the errors of previous efforts. This process is perpetually one in which the experimenters must make judgements between technical frameworks from which to proceed against a background of past efforts and frameworks. Thus experimental science must be an ongoing activity that is constantly open to change but does not pursue the objectness of its object-spheres or its object-areas, or even "Nature". The object-spheres and object-areas within distinct specializations in experimental physics are comprised of and structured by particular machine-kinds as their fundamental elements. Particular specializations are delimited by specific kinds of machine, and that novelty and new specializations arise from the innovative interconnection (the relating and conjoining) of machine-kinds. The object of the ontology of experimental physics is limited to machine performativity but, the function of this ontology is to provide a set of technological objects to disclose the mechanisms that are supposedly in operation during the "changing of the changeable". The phenomenal object of experimental physics is the machine performativity and that this object could only be conceived as natural by presupposing mechanical realism. On this account, the function of an object-sphere is to disclose the fundamental mechanisms that explain how human interventions within an object-sphere generate the machine performances actively produced and investigated by modern experimental physics. Modern experimental physics is only concerned with the motion of matter insofar as it discloses those fundamental mechanisms at work in a machine such as a pendulum or neutrino detector. It is only by recognising the metaphysical centrality of this concept within the rationality of physicists that we can understand the purpose of the performative art of experimentation and the nature of the reality that physicists aim to disclose. Thus we can see how physics has been performed for its own sake and bound up with *Ge-stell* in the making of the ground plan, the ongoing researches of all its fields of endeavour, and the construction of the theory of the real.

5
The Anvil of Practice and the Art of Experimentation

In this chapter I shall describe how experimental physics is a labour process of making and modelling the interactions between human interventions and machine performances. I shall describe how natural mechanisms are abstracted from an interpreted cluster of the alethic possibilities of labour that are creatively extended, interpreted, and manipulated by using analogical and metaphorical connections with clusters of theories, models, machine performances, and techniques. These analogical and metaphorical associations allow mechanisms to be transferred as technological objects across the boundaries between otherwise distinct experimental machines. Each prototype is a hybrid constructed through trial and error processes of bringing together and integrating heterogeneous technological objects from other distinct kinds of machines. This provides the innovation of new machines with a sense of developmental continuity and transfactual connections between machines that share component technological objects in their construction. The technological objects used in the innovation of each prototype have transfactuality in so far as they can be related through machine-kinds. When this stratum of new machines shares a common history of development then they are members of the same generative *machine-family*. The history of experimental physics is the history of the generation and extension of machine-families through the invention and innovation of prototypes using shared machine-kinds. Novel physics creates and explores unions between previously distinct machine-kinds and associated transformative powers. These are jux-taposed with the members of other machine-families when a novel hybrid is produced and the clusters are transferable throughout the machine-family along the genealogical lines of association. The ontology of physics is abstracted from these interconnected lineages of prototypes and their associated clusters. Highly complex machines, such as particle detectors, are composites of the members of several machine-kinds, and their distinct histories of development constitute machine-family in their own right and also show many overlaps with members of other machine-families. It is the transferability of clustered technological objects along the "genetic" lines of machine-families and their overlaps that gives these technological objects transfactuality.

Prototypes, such as Thompson's cathode-ray apparatus to measure the charge-to-mass ratio of cathode corpuscles and Millikan's oil drop apparatus to measure the quanta of electric charge were not as independent as Hacking and Bhaskar would like to believe. These experiments shared members of the same electromagnetic machine-kind in their construction, both utilised the same inscription practices, technographe, exoreferences, theories, and techniques, and were members of the same machine-family developed to determine the electrical properties of matter. Due to the assumption of mechanical realism, the technological framework is taken to be transparent as a means, and mathematical functions encoding change and persistence within machine performances are metaphorically presented as abstractions or representations of natural mechanisms. Through metaphorically based rhetorical, poetical, and visual techniques, emergent through socially mediated communication and argument, these abstractions can be verbalised, visualised, and conceptualised within models and theories of the causes in operation in the occurrence of natural phenomena. These mechanical models and theories are subsequently legitimated through a process of successfully implementing these abstractions as mathematical functions and exoreferences in the subsequent extension, innovation, and invention of their associated machine-families, and feeding them back into processes of guided technological innovation. An acceptance of mechanical realism has allowed this techneic innovation of the production of novel machines and mathematics to be epistemologically justified as an epistemic process of the discovery of natural mechanisms and laws. It is this process that provides a trajectory of stratification and correspondence between physics and "the strata of reality" that it discloses. This is how physics progresses.

It certainly is evident that the Glaser bubble chamber to detect "cosmic rays" and the LEP "antimatter-matter annihilating" ring at CERN are very different machines for particle detection. However, they do share important exoreferenced functives such as voltage, current, magnetic field strength, momentum, energy, etc., without any significant variance in usage; common elements, such as differential calculus, statistics, algebra, arithmetic, SI units, etc, as well as common mechanisms, such as radioactive decay, electron and photon interactions, electron and positron production and annihilation, spin and energy levels coupling, virtual particles, etc.; common indices such as charge, mass, spin, force, momentum, energy, half-life, etc.; both also utilise special relativity, Maxwell's equations, SI units, and the periodic table, etc., to provide a universal standard for description; and also use the same construals (e.g. particle, wave, track, spin, interaction, etc.) Furthermore, the computers are connected to the LEP-Ring detectors run reconstruction programmes that technographically represent the "data" as "bubble chamber" pictures of "tracks" on the computer screen and allow a commensurable visualization of the machine performativity. The LEP-Ring is a hybrid between the liquid-based detectors, such as the bubble chamber, and the gas-based detectors, such as the Geiger counter. These two prototypes were related through construals of their performativity in terms of the mechanism of ionization, and the particle detectors at CERN (DELPHI, OPAL, ALEPH, and L3) are, to put it crudely, not much more than ten metre by ten metre massive barrels of thousands of modified liquid and gas based detection cells, surrounded by a massive electromagnet and connected to the LEP-Ring. Each cell, when triggered, transmits a voltage peak, a time signal, and an ID number. That's all! Computers, reconstruction techniques, and models do the rest. The use of the "ionization" construal brings together and connects chemical and electromagnetic machine-kinds, making the bubble chamber (cloud chamber, drift chamber, etc.,) and the Geiger counter (xenon tube, arc lamp, etc.), the detectors at CERN, and the LEP-Ring members of the same machine-family. They are linked historically and technologically as innovations of the same project (i.e. the physics of the motion of interactive charged particles in electromagnetic fields through liquids and gases), that was projected using the same ground plan (mapping out the connections between the geometry of kinds of events with the material conditions of those events), using permutations of the same machine-kinds. It is this transfactual membership that is metaphysically conceptualised by the mechanical realist as a process of disclosing stratified ontological depth when the development of any machine-family is made transparent as a mere means. Mechanical realism allowed the historical innovation and development of a machine-family to be conceived as the process of following the logic of the stratification of Nature.

The frontiers of new experimental physics are the new challenges of *Ge-stell* and the already stabilised technological objects of prior efforts achieve their concrete reality and transparency to the extent that they are available as standing-reserve. In this way, the destining of *Ge-stell*, as the destining of physics, promises novel technological complexes of objects that will challenge the labour processes of physicists to disclose deeper ontological strata through innovation. The frontier of physics attains an ontological "superiority" over the past efforts that its destining is entirely dependent upon. This order of rank is entirely one of the "superiority" of challenges over standing-reserve. It is an expression of the desire for novel experiences and powers. The new strata of machine-kinds, historically situated within a machine-family, is the "cutting-edge" or "state of the art" of novel physics that is manifest as a research programme through its subsequent innovation into further differentiated machines. The ontology of experimental High Energy Physics is circumscribed by the permutations of particle detectors and accelerators innovated by the extension and synthesis of the same component machine-kinds. Hence, the machine performances of the LEP ring accelerator at CERN and the linear accelerator at SLAC in the United States are represented as disclosures of the interactions between fundamental "particles" and commensurable means to test and refine the same models and theories. This possibility is a consequence of the fact that the same models and theories were built into both machines through the processes of stabilising the convergence and connection of different permutations of the same machine-kinds during the design and construction of those machines. It does not follow from the fact that these machines are different that they are independent. There are merely different ensembles, different configurations, of the same component technological objects and, as such, are members of the same machine-family. These machines are taken to be independent, or autonomous, because their shared component machine-kinds have become transparent as standing-reserve. Any *technai* (as an ideal) disclosed through experimental physics are represented as "natural laws", whilst the modes of disclosure entities such as electrons, for instance, are limited to their instrumentality within interconnected strata of machines for which electrons are utilised in the inscription and interpretation of their performativity.

Models, metaphors, and machine performances

The modelling of the performativity of machines, in relation to the labour processes of the research programmes that circumscribe them, is a process of positing causal relations and mechanisms that are relevant to the particular goals of that research programme. The transference of these models between members of the same machine-family, as a process of cross-checking and modifying both the models and the performativity of the machines, transforms the posited causal relations into a more abstract code and leads to their generalization. The experiences of the research workers on one particular machine can be related to the experiences of the research workers on another particular machine, through these generalised models, unifying these machines as members of the same machine-family. The physicists working on DESY in Hamburg, the DELPHI detector at CERN, and the SLAC machine in the United States, for example, can relate their experiences through the abstract and codified Standard Model of Elementary Particle Interactions. This occurs despite the fact that this general model needs to be extensively expanded, refined, and modified to be of any use in exoframing their particular projects. The physicists are able to generalise and unify their experiences as experiences of the same kind of events by relating their experiences in terms of a shared general model, and, as a consequence, transcend the particularity of the specific machine performances. It is only in relation to a shared general model that these physicists are able to translate their particular experiences of the particular performances of particular machines into general observations of the same processes. This endows those events with autonomy and transfactuality. Transfactuality should not be of any surprise given that it was built into the machines from the onset and has been continually maintained by the experimenters in their interpretations of their experiences. The transfactuality of those experiences is a consequence of the transfactuality of the exchange of experiences in terms of a shared general model that is both built into the design of the particular machines and is used to interpret and exorefference their performances. Without a general model, those experiences would be incommensurable because they would lack any common frame of reference, and, therefore, the use of a general model is a precondition for any division of labour within ongoing experimental research programmes. The use of a general model is also a precondition for the subdivision of any research programme into particular experiments and further specialization. The use of general models also allows the physicist to de-personalise their experiences and represent them as the experiences that anyone would have (providing that they were familiar with the general model). It is de-personalization of the final accounts that is represented as objectivity and facilitates the removal of the particular labour processes that were necessary for the physicist to have those experiences. This is an example of reification. The experiments and theories of modern physics, presented as autonomous and transfactual in terms of a general model, can only be divorced from the challenges of *Ge-stell* and the teleological positings of the labour process by presuming the validity of the precepts of mechanical realist metaphysics. This is an expression of the public conflation of *techne* and *episteme* that has been central to experimental physics since its onset in the work of Galileo. The general causal accounts of the stable products of the socio-technical processes of labour can be presented as universal knowledge of the eternal and fixed efficient causes of Nature in terms of the general mathematical laws that describe the mechanisms responsible for phenomenal changes. It is at that point that both human agency and machine performativity have been written out of the account completely. This conceals the teleological positing that is inherent in the destining of experimental physics because, by presenting the directions and products of physics as the disclosure of natural processes, it has hidden the purposes and challenges for which the experiments were set-up to satisfy. This concealment hides the beings for which the disclosures of physics are truths and powers, and, as a consequence, not only hides the choices that have been made regarding the human inquiry into Nature, but also hides the fundamental relation to Being that provides the precondition for those choices. It is as a result of this concealment that *Ge-stell* became the natural mode for being-in-the-world, and, simultaneously, became concealed from presence as *Ge-stell*.

A model of any real process is inherently a simplification, abstraction, and an approximation of that process. In experimental physics, the labour processes of making a model involve abstracting the mechanisms and functions of the apparatus in order to make manipulation of the model as simple as is pragmatically acceptable whilst simultaneously making selections as to which features of the apparatus' configuration are essential within the particularities of the research project. The purpose of a model is to represent reality and to simplify it. Physicists, when constructing models, have to make a choice between making the model simpler (easier to work with) or making it more complex (realistic). Physicists abstract, picture, and generalise a real process, or the world, and only the most naive empiricist would claim that a perfect model perfectly represents reality. If a model were to mimic reality in every detail then the model would be useless. Furthermore, the purpose of a model is to explain a real process (by relating structures, mechanisms, or functions to objects, phenomena, or systems) in order to explain the various emergent properties that it has. Models allow the processes of experimentation to continue through relating features of phenomena, experimental procedures, and exoreferences to "physical variables". A model is a tentative, re-evaluative pragmatic tool evaluated according to its instrumentality in making descriptions and explanations useful and intelligible. It is an essential link between mathematical functions and experiment, providing the exoreferences needed to connect the machine performances of exoframed apparatus and calibrated instrumentation with a "physical process". Exoreferenced mathematical expressions can be used as functives within exoframed apparatus that relate the calibrated instruments and machine performativity of apparatus with the "physical processes" that are being experimented upon by providing a visual, descriptive, explanatory, and intelligible representation of the "changing of the changeable" of the experiment. They are used to design and construct instruments, experiments, and computer simulations, as well as to calibrate measuring devices, and interpret experiences in novel contexts. Models are transferable across contexts in which a common abstract feature between those contexts allows the implementation of the same abstract model. This allowed Laplace's equations, for example, to serve as a mathematical model for quantifiable change in diverse fields such as gravitation, electrostatics, electricity, elasticity, and liquid flow. It is the use of such models as analogies that unifies and informs otherwise distinct fields of scientific research.

Francis Bacon regarded the use of analogy as essential to scientific thinking because the investigation and observation of resemblance and analogies provides us with a sense of the unity of Nature and a foundation for scientific inquiry.[1] Analogy is a fundamental technique for enhancing intelligibility and facilitating communication by comparing the phenomenon to something familiar and to be drawn in terms of visual representations. As Gooding argued, it is via using analogies such as "curves of force" – one of the central visualizations in electromagnetic theory – that Faraday was able to visualise the phenomena under investigation and construct models and develop novel experimental techniques, models, and communicable experiences.[2] There are many other examples of analogical reasoning and modelling in the history of science. Galileo used the Jovian satellite orbits as an analogical support for Copernicus' heliocentric solar system. Kepler used an analogy between musical harmonics and planetary orbital geometry. Newton used a terrestrial projectile as an analogy for the Moon. Bohr used the heliocentric solar system as an analogy for the hydrogen atom. The free electron theory of metals, the kinetic theory of ideal gases, the Ising theory of ferromagnetism, the Standard Model of Elementary Particle Interactions, the Big Bang theory, etc., are all examples of analogies used to construct intelligible models to explain phenomena. This analogical feature of models is essential for the heuristic of scientific discovery to occur at all by providing explanations of novel phenomena in terms of familiar narrative and imagery. Through verbalization and visualization the novel can be related to the familiar in such as way as to suggest familiar techniques to innovatively explore the novel. In this way a model can be used to make sense out of a phenomenon by relating it to things that are already pragmatically understood. An example of using a model in this way would be the wave theory of light, which was demonstrated using ripples in water to explain phenomena such as refraction and diffraction. The use of such a model provided physicists, such as Young and Fresnel, with transferable techniques by which they could build an apparatus to demonstrate the wave-like nature of light. The analogous use of the wave theory of light to exoreference the wave-functions of Schroedinger's equation, or QED integrals, to observational events in the physical context of the double-slit experiment with "electrons", in order to disclose "wave-particle duality", shows how the analogy with the Young and Fresnel experiment permitted probability theory to be postulated as an intelligible model of the "unobservable" quantum world. The analogous relation of the technical definition of "probability wave" in mathematical theory with water waves communicates and represents the perceptual phenomena to the machine performance in order to make the experiments and mathematics intelligible as means of disclosure. The heuristic use of the wave theory of probabilities to interconnect logical, mathematical, and computational analysis of "sub-atomic phenomena" provides a technological framework in which calculations and observations can be made and also a logical structure through which "thought-experiments" can be constructed. This allows a model to be used to make predictions by calculating numerical values, in relation to any variation of factors, and also to derive new hypotheses, theoretical conceptions, and possible observations.

Morpurgo's use of Millikan's oil-drop experiment as a model template for a novel apparatus to search for free-quarks shows how the analogical use of a model can also suggest ways to associate "well understood" experiments with new experiments.[3] The apparatus was an analogical association between otherwise distinct theoretical entities. This was a bounded technically rational strategy because, by analogy, what Millikan had already established about discrete electronic charge using his oil-drop apparatus, Morpurgo could potentially establish about discrete fractional electronic charge (the theoretical characteristic of quarks), and consequently use Millikan's apparatus as a model for an apparatus to search for free-quarks. At each and every stage of construction, his choices of components and their interconnection were analogously situated in a pre-existing technological framework by using a model to refine and develop that framework. Techniques for the identification, measurement, and manipulation of "charge" could be transferred, by analogical use of a model because both machines shared components that were members of the electromagnetic machine-kind, such as capacitors and electromagnets, and, through the use of transfactual indices and functives, such as "charge", in the context of shared machine-kinds, they could use commensurable exoreferences to interpret their performances. This shows how analogy plays a synthetic role in bringing together heterogeneous domains of experience and allowing them to cross over boundaries between distinct specializations and bring them together as members of the same machine-family. Throughout the history of physics, new models often emerge when two previously distinct disciplines or fields are brought together. For example, cosmologists trying to understand "dark matter" have sought a mechanism by which it can be detected. In contemporary quantum physics, "superfluidity" (itself an analogy) can be taken to be an analogy for "a vacuum" in which changes in "the AB-boundary position", between "the A-phase" and "the B-phase" of "superfluid He-3", can taken to be analogous to "the symmetry breaking" that "occurs because of certain cosmological strings of dark matter" that are "predicted" by "the Inflationary Phase of the Big Bang" model (based on an analogy between space-time and an expanding gas).[4] It is because of analogical reasoning like this, that machines such as dilution refrigerators (previously used to map out the properties of superfluid He-3) can be used as possible "dark matter" detectors and the two previously distinct ultra-low temperature quantum physics and theoretical relativistic cosmology can be brought together.

An explanatory model can only be said to approximate the truth and maintains its "ontological realism" through the commitment of its adherents and allies if it is an intelligible model. If one postulates that light was really comprised of waves then the wave theory of light has been mathematically projected over the phenomenon of light in order to make the behaviour of light intelligible. The physicists skip over the phenomenon and interact with the model as if it were the phenomenon. An example of this is the demonstrative and educational use of electrostatic machines in the eighteenth century to suggest a connection between electricity and lightning postulated a heuristic connection between such machines and the means to explore the natural phenomenon.[5] Of course, it is not necessarily the case that a model is considered in this way. The instrumental and metaphorical use of models allows physicists to explore the postulated ontology whilst remaining tentatively sceptical about the final truth-status of the theories that they use. Such models are used instrumentally as a technique and an inspiration. Many physicists use quantum mechanics in this way. It facilitates calculations and predictions but the physicists do not necessarily commit themselves to its reality. However, in an ongoing physics experiment, the apparatus is itself constructed according to models of the performativity, how those components will relate to one another, and how they will interact with the phenomenon under investigation. Many such models built into the apparatus in its set-up to relate its components and realist interpretations of these models are necessary if the apparatus is to operate as a means of disclosure. By using models to accommodate complicated phenomena within the technological background, physicists are able to use a model to provide intelligibility, cognition, and articulation, whilst relating variables to model the apparatus in terms of functional sequences and relations, which are connectable to operations and procedures. Modifiable exoframes have a capacity to accommodate development and change in the configuration of the apparatus and the understanding of the phenomena, and to accommodate discrepancies and problems that arise in their use. This allows for the quantification of properties (an essential factor for measurement) and represents the phenomenon as a set of mathematical relations between functions, variables, and constants that are open to refinement and increased predictive accuracy. Epistemic and *techneic* uses of models are interconnected via exoframing, in order to model the complete physical experiment as a mode of disclosure by connecting the manipulative procedures of experimentation with "physical variables", and blur the distinction between representations and interventions. Both the configurations of the exoframed apparatus and the current model are refined and related to the disclosed techno-phenomena in order to establish correspondence between them on the premise that a stable and coherent procedure "must have some relation to the truth". The machine performances, of course, are eventually predictable using the latest refinement of the exoframe. Then the physicists publish.

Novel experimental physics is not simply hypothesis testing or theory falsification; it is a technological and theoretical ongoing process in which observations, models, expectations, and techniques are developed simultaneously in the context of making the experiment work as an experiment. Modelling, achieved through the connection of visual representations and mathematical functions, allows a model to act both technically and metaphorically by simultaneously modelling how something works and what something is. Thus the process of modelling is a two-fold process within the context of developing stable experiences and stable technical procedures, with the aim of constructing stable communicable observations. This process involves a reiterative interplay between the technical practices, instrumental modelling, and phenomenal modelling, which interact with each other as the working experimenters attempt to construct stable scientific solutions as perceptible and intelligible to both the experimenters and the wider public. Models allow for human imagination, reasoning, argumentation, negotiation, visualization, and intuition, to be active in bringing "a natural mechanism" into the public realm, by making "it" and object for manipulation, cognition, and conception in terms of familiar objects and relations. This use of modelling connects perception, imagination, and intuition, with "natural mechanisms" via technical manipulations and procedures. It allows the experiences of physicists to be translated into different contexts of ongoing and innovative material practices, and the success of an analogy in scientific work is rhetorically supported by reference to its usefulness rather than its truth, on the basis that "it is a model which works". It remains techneic, situated in the context of developing and tentatively evolving practices, apparatus configurations, and skills, but by virtue of its instrumental success, it is treated as if it has some probabilistic epistemic correspondence. Probabilistic arguments require attendance to already established beliefs, assumptions, prejudices, and opinions, if the arguer is to be successful in establishing the analogy as probably true, because they are attempts to establish a similarity, a likeness, in terms of the familiar and already accepted.[6] In order to secure epistemic correspondence, a model must be rhetorically secured to the already established conventions of a community. This involves using imagery, metaphors, assumptions, predispositions, and values, which the community already has, in order to present the enthymeme as convergent and coherent with the community's conventions. By cohering with already established conventions, the argument has the character of a "naturalistic" argument; the community is more likely to accept what is possibly true as probably true, and what is probably true as true. The success at achieving this depends on perceptions of credibility, appropriateness of the analogy, the inspirational quality of the analogy used, utility of the model, and the difficulty (and risk) of refuting the analogy. This requires widespread distribution through credible media and also the refutation and avoidance of any criticism. When making choices regarding what to put into a publication, and what not to, scientists already pre-empt possible criticisms to their results. This means that scientists are engaged in the public acceptability of their work when making their choices about the direction of that work. In Kuhn's terms, such an argument must be made within a shared paradigm and there would be a set of exemplar arguments that would have a high chance of working.

A model can be constructed out of visual, verbal, and mathematical analogies but, in order to qualify as a legitimate model in experimental physics, it has to facilitate the derivation of measurable quantities. This constraint means that the possibilities available for any particular model are limited by the physicists' expectations of measurability. This expectation is constructed according to expectations of cost, in terms of credibility risk and available resources, and perceptions of the limitations of available technology. The expectation of what is measurable is decided in reference to the already established labour processes, the economic support available to the group, and the conventions of measurement. This expectation of measurability situates the available choices of models to be consensually commensurable within already established practices of the technological framework. It is in relation to this axis of commensurability that the perception of measurability and manipulability is constructed as a perceived functionality of the available machines. It is this perceived functionality that acts as a constraint upon theory and practice. "Natural phenomena" in experiments are *reduced* to exoframed machine performances and, therefore, the object of experimental investigation is itself an analogy, a model, of the "natural phenomena" under investigation. This show how there is a "gap" between the experiment and the natural world that can only be "bridged" using metaphysically underwritten metaphors. Empirical experience is defined within experimentation, via a technological constraint upon the form of any models used and what the possibilities of use could be, which reduces what the form of a scientifically demonstrable ontology could be. Physicists will not consider any ontological description to be empirical unless that description has been produced within and validated by the technological context of experimentation.

This technological framework extends to visualizations, interpretations, conceptualization, and mathematical abstractions. A model is invalidated if it cannot be subjected to experiment and it must remain within the boundaries of manipulations and demonstrable context using publicly acceptable techniques. Thus the natural properties associated with "copper", for example, are those, and only those, properties that can be mechanised and repeated within a technological framework. When these properties are taken *to be* the "natural phenomena" the experiment replaces them with model constituted in terms of a set of exoframed mathematical functions, stable interpretations, and related visualizations. By doing this, the physicists are simultaneously taking the model literally as the phenomena and metaphorically by treating the technologically constituted model as the natural phenomena. The model has replaced the phenomena through mathematical projection, in Heidegger's sense, whilst being treated as a mere instrument. The situation in experimental quantum physics is even more ambiguous because we are in a situation where our only experience of the phenomena occurs in the laboratory. There are no experiences of entities such as "superfluids", "neutrino oscillations", or "tau-leptons" that occur outside the highly technological environments of the production, detection, and modelling context. As such these machines do not straightforwardly constitute metaphors because they do not replace the phenomena but constitute the context of the emergence of the phenomena. This physics has the reverse situation where the "natural phenomena" are metaphors for the machine. The machine is replaced by "natural phenomena" and as such the latter is a metaphor for the former.

A metaphor is usually defined in terms of a "deviation or displacement from literal meaning". However, a notion of "literal meaning" is extremely difficult to define and maintain except by appealing to notions of conventional usage and context. This is even more problematical when it is pointed out that the definition of metaphor in terms of deviation or displacement of the meaning of a word from its literal usage is itself a metaphor because the meaning of words cannot literally move.[7] The ambiguity of metaphorical usage is heightened when comparing models and phenomena because we are not exactly making a literal correspondence between them but rather superimposing one over the other. A metaphor preserves the sense of the model not being the phenomena whilst being taken to be interchangeable with the phenomena. It involves equating the previously unequal whilst preserving the sense of their inequality and simultaneously treating one as the other.[8] By using a model as a metaphor, physicists are able to treat the model as the phenomenon, replacing the phenomenon with the model, whilst distancing themselves from any commitment to that equation by maintaining that their model "is just a model" and is not identical to the phenomenon. In this sense, modern physics does not differ from poetry in that it generates new ways of seeing aspects of the world through metaphorically facilitating a transformation of convention, via a sense of appropriateness and intelligibility, in its search for fundamental truths. The use of models as metaphors brings novelty and invention into experimental physics by allowing transactions between distinct contexts of material practice to transfer metaphorically across machine-kind boundaries by using models from analogous projects as templates. It allows them to tentatively explore the possibilities and actualities and significance of the experiment, without any certain knowledge, utilising the novelty and invention of the model metaphorically in order to secure it as a plausible approach to making the phenomena intelligible. It also allows the possibility of novel machine hybrids. It does this by bringing together heterogeneous technological objects within the technological framework of its metaphysical performance art.

Experimental physics is a techno-poetics in which the models, as metaphors of experimental imagination, are physically embodied, through material practice, in the mapping out of the contours of the interactions between human interventions and machine performances. It is akin to making an automaton from other automata and situating it within a world-picture. If we consider the case of the physicists' use of "superfluids" to model "cosmic strings" then we can readily see the techno-poetics of experimentation. What is the metaphor here? It is through the juxtaposition of two techno-poetical metaphors, the quantum fluids and those of cosmological topologies in the inflating manifold of the quantum flux of space-time, that these metaphors are technically operational as a novel search for truth. Two previously distinct clusters of techno-phenomena are reflected against each other and one is presented as a model of the other in order to illuminate both and also create novel meanings. The metaphor is the substitution of the exoframed contours of the ultra-low temperature quantum machines' performances for the theoretically simulated exoreferences of dark matter cosmology. It is in this sense that experimental physics is a poetical and metaphysical performative art that substitutes *Ge-stell* for the quest for novel truths and experiences. This metaphor transforms the aloof world of cosmological technographics and exoreferences into the cluster of technological objects available to the ultra-low temperature physicist. The transdiction of invisible dark matter remnants of creation (something beyond perceptual acquaintance or "direct" experience, but used as a corrective to explain the existence of galaxies) come ready to hand through the techniques of ultra-low temperature quantum physics. Such a project is beyond the "rigors" of positivism! The experiences of digital outputs and technographic computer simulations of the performance of ultra-low temperature cosmological experiments are techno-poetical disclosures of events that allegedly occurred billions of years before the Sun was born and, allegedly would be evaporated, without trace, by the sizzling heat of a snowflake. There is nothing immediately sensory about such disclosures. They are transdictions designed to solve the cosmological absence of the theoretically required number of visible galaxies, explained in terms of the resonance signals from vibrating wires inside a dilution refrigerator. The "mechanisms" disclosed through these experiments are substituted as proposed mechanisms in operation to explain the machine performances of their instrumentation and computer outputs. The vibrations of the thin wires submerged in the near nothingness of superfluid He-3 are techno-poetical transdictions as a form of "reverse engineering" which reconstructs machine performativity in terms of the invisible interactions between "quasi-particles", "cooper-pairs", and "energy gaps". It is then used as a working model of the motion of the dark matter remnants of creation within the unfolding manifold of space-time. The "invisible" is made visible through the metaphorical substitution of another "invisible". How could we not see the poetry of this?

As producers of metaphors, physicists are directing their creativity towards making aspects of the world intelligible. They may well rhetorically situate these new ways of seeing as objectively corresponding to something "out there" on the basis of their claim that they are genuinely engaged in attempting to make parts of the world intelligible in novel and interesting ways. However, what they have missed, through familiarity, is that these parts of the world are made and have transformed the way that the physicists relate to the world. In this respect physicists are engaged in a very human pursuit that is located, anchored, and directed from within a world-picture. All theoretical cognition takes its departure from a background of language use and material practice that precedes the theoretician. It was only in virtue of being anchored, located, and directed, within a cultural background that physics has been able to create and disseminate the poetical "mechanical world-picture". The development of novel physics is bound up with a mode of being-in-the-world that cannot be divorced from the cultural background that makes that mode possible. This cultural background gives modern physics its reality. The metaphorical use of pictures and technological objects is an extension of this cultural background and, as a consequence, it is a socially constructed reality. Metaphors allow a disordering and ordering of social imagination in such a way as to say something about one thing in terms of another. By allowing selections, emphases, suppressions, reductions, and organizations, of the components of novel subject matters to be made in terms of other subject matters, metaphors are essential to the development of new ideas and are not merely "dispensable graces and ornamentation."[9] Metaphors are essential for the development of novel technical languages from established technical languages and ordinary language.[10] However, when we begin to lose the sense that what has been made equal is not equal, when we loose the sense the "model" and "phenomena" are not the same, then our metaphors cease to be metaphors and become literal and assertive truth claims. This strategy uses rhetoric to move from poetics to establish something as plausibly true in terms of the current conventions, current perceptions, beliefs, standards, values, presumptions, dispositions, assumptions, prejudices, etc. The transformation of the metaphorical innovation of language into literal usage is a transformation of the background of conventional language usage available for future transformation. Physics is a mode of agency directed towards the innovative production of models and technological objects by using metaphors to transform and extend the reality that it is making. This involves the poetical and rhetorical use of metaphors in constructing pictures of Nature and also the innovative use of technological objects metaphorically across contexts of production. The reality of physics is brought forth into Being. It is in this sense that physics participates in *poiesis* and "world-making", as well as in the sense of providing the world with world-changing prototype machines (such as the electric motor and the atomic bomb) which the physicists "brought into being". The nature of *poiesis* has been transformed by the performative and productive art of experimental physics to explore truth in terms of a series of disclosures of our productive and performative possibilities. This kind of truth does not readily support either realist or positivistic interpretations of physics. As Nelson Goodman put it,

> What I have said so far plainly points to a radical relativism; but severe constraints are imposed. Willingness to accept countless

alternative true or right world-versions does not mean that everything goes, that tall stories are as good as short ones, that truths are
no longer distinguished from falsehoods, but that truth must be
otherwise conceived than as correspondence with a ready-made
world. Though we make worlds by making versions, we no more
make a world by putting symbols together at random than a carpenter makes a chair by putting pieces of wood together at random.[11]

Truth cannot be divorced from the beings for which it is a truth
without making it unintelligible. It arises as truth through a mode of
agency and is disclosed through making it into a disclosure. It is
brought-forth as truth. It is a mode of *aletheia* and discloses the way
that experimental physics belongs to bringing-forth and *techne*. The
establishment of realist belief occurs when models are taken literally
instead of metaphorically. This substitutes a model as a phenomenon.
The *techne* induced and abstracted from the ongoing productive practices of experimentation is disclosed as *episteme*. This facilitates the
removal of the mode of disclosure from the context of working practices and social agents through which it occurred. Once this has been
achieved then its products can be represented as "abstract general principles of how Nature works" and represented as yet another confirmation of the success of physics. It is through taking literally the
metaphorical substitution of mechanistic models "of Nature" for the
mechanistic *techneic* modelling process that allows the machine performances and manipulative procedures of the experiment to be dropped
from physicists' accounts and replaced with abstract mathematical
"Natural Laws" and causal mechanisms. It is due to this process of
reification and abstraction that experiments can be seen to test
hypotheses, falsify theories, and that the phenomena in question are
represented as the product of a set of "natural mechanisms". The
metaphor of "natural mechanism" has been established in our culture
for at least 400 years and is the cornerstone of experimental physics.
The establishment of experimental physics was simultaneously the
rhetorical establishment of the literalness of "natural mechanism" and
the transparency of experimentation as a means of disclosure. The
establishment of the method of experimentation as a road to truth was
only possible because of the metaphysical precepts of mechanical
realism. Techniques and machines could be treated as transparent
means to the truth about natural mechanisms. The discourse of physicists could be then presented as directly reading the Book of Nature
rather than writing it. Innovative modelling, developed in the process

of making the work of making material practices intelligible, is a metaphorical process that is used rhetorically and poetically to disclose the discovery of human productive and performative abilities metaphorically as a disclosure of natural law. As Ralph Waldo Emerson put it,

> Man is a shrewd inventor, and is ever taking the hint of a new machine from his own structure, adapting some secret of his own anatomy in iron, wood, and leather, to some required function of the work of the world.[12]

The technological framework of experimental physics

As Nicholas Maxwell pointed out, we must not forget what sort of being scientific theories and models are intelligible for.[13] Once we address intelligibility as a central criterion for theory and model choice then we cannot remove the living bodily being, namely scientifically persuaded human beings, for whom those theories and models are intelligible, without removing their intelligibility. Furthermore, as the choice of theory or model is bound up with the choice of research programme, then we can not detach the question of which theories and models are to be explored from the teleological positings in the "external" world of commercial, political, and military ambitions, that the direction of research has promised to satisfy in exchange for resources, equipment, and space. Modern experimental physics is situated between two dimensions of productive activity: the "internal" dimension of the production of intelligible explanations of machine performativity in terms of natural laws and mechanisms, and the "external" dimension of the production of innovative transformative powers and machine prototypes. In order to understand how these explanations of machine performativity are made intelligible and communicated then we need to examine the complicated processes involved in learning how to have experiences of otherwise "invisible" processes and entities. This involves understanding how someone becomes a skilled experimenter.

Any new research student entering a modern experimental physics laboratory for the first time enters a highly complicated technological environment populated by already established practitioners. S/he is a novice despite her/his familiarity with technology use and science. Her/his previous experiences and acquired knowledge of basic techniques and scientific theories are insufficient for the purpose of making this new environment intelligible. S/he requires further training. Many students and researchers anecdotally recall being told to "forget everything they learned before" as they are initiated into "the world of research". Novice students frequently spend the first year "familiarising themselves" with the experimental apparatus, techniques, procedures, theoretical models, computer simulations, and "the way things are done" within the laboratory. They learn the techniques and tactics. They are also expected to familiarise themselves with "the current literature" and "who's who in the field" as the definition of "the field" and "the state of the art". They also learn what the laboratory's projects and aims are, as well as who are their allies and competitors. In short, the novice student is socialised as a competent laboratory member by being orientated within this specialised technological and social environment. A novice student's prior familiarity with using technologies and her/his previous knowledge can only help as a starting point. For the purpose of participating in the specialist character of all scientific research, the idiosyncratic character of particular laboratories' working practices, and the novelty and complexity of the experimental research, the student's previous education is too general, too basic, and also often obsolete. The student's education would not provide the level of refinement of skills and technological familiarity required when working upon the experimental apparatus.

Although research physics students often attend general theoretical courses, learning both general mathematical and theoretical techniques, most of the training will be in the context of the specific laboratory work. The novice student needs to rapidly make the highly complicated technological environment intelligible through the embodiment of the already established laboratory practices. S/he learns the methodology. The first stage of this process of orientation requires a great deal of "black-box" abstraction of the laboratory's activities and the experimental apparatus into sequences and series of functions, operations, and procedures. These range from the operations of devices and instruments in specified circumstances to the procedures for recording operations in laboratory notebooks. The novice is taught specific bodily acts and technical operations; this orientates the novice as an experimental practitioner. This form of learning is entirely one of attendance on the part of the student, and instruction (both formal and informal) on the part of the already established practitioners. The term "attendance" denotes certain attitudes on the part of the student. This term means more than simply being present. It denotes an attitude of *listening, attentiveness,* and *care*. It is used in the same sense as when someone attends to their duties or attends to someone else's needs. This term implies that the student is engaged in more than a passive learning relationship within the laboratory. The student is actively orientating him/herself through making his/her working environment and the project(s), in which s/he is a participant, both intelligible and her/his own. It is through this habituation and familiarization, in terms of "when this gauge reads A then turn that dial to B because this performs function C" instructions, that the technological environment is embodied in practices as a set of techniques. This socialises the student into "how things are done" in terms of operational procedures that are represented in terms of cause-effect sequences. In this way, the novice learns how to operate the experimental apparatus and become a competent practitioner. The novice student also learns, as an equally important part of learning "how things are done", the social dynamics within the organization of the laboratory, and how to orientate her/himself within everyday working practices and relations. This involves learning "who is best at what", "how to approach so-and-so" in order to obtain their help, how work tasks are organised, how meetings are organised and resolved, and even how to join in with the laboratory members' "sense of humour". In short, the novice must learn how to position her/himself within a labour process. The social dimension of the labour process is as equally important as the technical operation of the apparatus for the student's acquisition of competency. The student needs to be orientated into the socio-technological organization of the laboratory in such a way as to become a negotiator rather than merely instructed, and must survive a process of socialization through which the student's relationships are both dialectical and didactical. The didactical aspects of the student's relationships are those in which s/he learns established practices through following instructions and mimicry. The dialectical aspects are those in which the student negotiates her/his own position in the labour process and the experimental apparatus through questioning and participation.

However, the *praktognosic* processes of the embodiment of techniques are necessary but insufficient to make the laboratory environment intelligible and workable. The empirical constant conjunctions between machine performances and the interventional techniques are exoframed, from the onset, with theoretical descriptions and representations of "invisible physical processes". The student learns how to use models to exoreference techniques in terms of "physical processes" as an essential part of making the process of experimentation intelligible. This is how the student acquires the necessary background to be able to have these experiences for her/himself. It is a process of *mathesis* emerging from the praktognosic embodiment of the exoframed technological framework. The student is taught how to make the performance of the experiment intelligible in terms of interpretive models, mathematical laws, theoretical conceptualizations, and visual representations of invisible "physical processes" that are claimed to be occurring within the experimental closed system. For example, a physics student might learn that "adjusting the yellow valve marked B changes the pressure of the helium-3 flow", that "switching on circuit A turns on the magnetic field", or that when digital display C reads "D" then "the phonon absorption rate has reached a significant level". The "physical properties" then can be "actualised" by pressing switches, or turning knobs, which tacitly relate techniques to "adjusting energies", or "quantum states", or "temperature", or "magnetic polarization", etc. This tacit juxtaposition between procedures and the manipulation of "physical variables" is an interpretive, technical, and cognitive orientation within the process of making the laboratory intelligible. Embodying this tacit connectivity in practice allows the novice student to participate in "doing physics". *Mathesis* makes the abstract level of theoretical physics manifest in, and linked to, the material practices of the operation of the apparatus and the student's experiences of the performativity of the experiment. S/he learns how to participate within the labour process and how to make sense of the products of this process in terms of experimenting upon the correspondence between models and "a physical process". By learning how to use models to interpret the performances of the apparatus, the student learns how to conduct scientific negotiation and reasoning, and becomes an experimenter. The extent to which the student will be perceived by the already established practitioners to have become "competent" will be dependent upon the student's success in orientating her/himself within the context of the labour process as a productive practitioner. It is by doing this that the other practitioners will perceive the student as having acquired the skills necessary to do the work. The student will become competent by embodying the methodology of physics and mathematical projection in practice, and by learning how to embody that methodology for her/himself in such a way that the connection between models and procedures becomes tacit. The student learns how to take measurements and make observations by implicitly relating exoreferences and techno-phenomena with the functionality of apparatus, which, in turn, becomes an increasingly transparent means of disclosure of theoretical mechanisms.

Experimental physics is a highly complicated and "messy" affair. The novice student has to learn not only how to make skilled judgements about how to proceed with the work, but also what constitutes a "skilled judgement" in the first instance. Experimental apparatus do not always work in accordance with expectations. In fact, many experimentalists often joke that experiments rarely work in accordance with expectations. The novice student learns the intentions and expectations of the laboratory in the context of productivity despite frequent instability and complexity. S/he learns what constitutes "working well" as well as "what to do", "how to do it", and "when to do it", as established tactics for dealing with the instabilities and complexities in their work as they attempt to stabilise productive working practices. In this way s/he acquires the received wisdom of "how to perform experimental physics" and "what the results should look like" if the experiment is "working properly". The embodiment of "the best way to proceed" to "achieve realistic goals" is passed on to the student through this orientation of the student's expectations and practices in line with the established and stable expectations and practices of the laboratory. The student also learns, in parallel to "working well", what constitutes a "breakdown". S/he learns "likely causes" of "breakdown", the "whys", and "the best way to fix it", as part of the process of familiarization with the laboratory's embodied history of trial and error experiences, modifications, interpretations, and goals. The "best way to fix this" is often a particular solution to a "failure" which is pragmatically taken to be the solution provided it remains stable. For example, the student may well learn that if a particular characteristic signal appears on a spectrograph and is considered by the already established group members to be "noise", and consequently "undesirable", then it should be "fixed", or "removed", by "changing a connecting cable", "adjusting the signal amplification", or whatever. Much of experimental physics involves the acquisition of a collection of experiences of tinkering with the technological configuration in order to maintain a stable output in line with expectations.

It is only when such attempts persistently "fail" and that the output persistently "fails" to meet expectations that the experimental practitioners will look "deeper". This "deeper" look involves making "a skilled judgement" regarding whether there is a serious problem with the construction of the experimental apparatus, the performance of particular procedures, or there is "new physics". This judgement is performed by running through a list of possible transdictions, either removing possible sources (until it vanishes), or exploring it through subsequent innovation (making it instrumental in the performance of its own disclosure). If the expected output can be achieved and stabilised by re-building part, or all, of the experimental apparatus then the problem is determined to be one of apparatus design or construction. The experimental apparatus was "flawed". If the expected output can be achieved and stabilised by using a different technique, or simply by someone else performing it, then the problem is determined to be simply a matter of practice. The experiment was "performed badly". If the expected results cannot be achieved through changing the apparatus design or operation then (typically only then) "new physics" will be suspected. Something was "misunderstood". Only if the anomaly remains stable and an impediment to the research project will it be considered worthy of further investigation. The extent to which Popper's "falsification principle" is a principle in scientific work is that, in principle, any scientific theory or hypothesis should be falsifiable, but working physicists rarely spend any time falsifying hypotheses or theories. Most scientific endeavours are attempts to make things work. A theory or hypothesis will only be rejected if it consistently fails to work and there is a workable alternative. Otherwise it will be avoided and the work will continue as before.

Anomalies are generally only seriously considered when the current model persistently fails and an alternative hypothesis is presented to account for them and suggest a new direction for experimental research. Without this "explanation" the experimental result rarely is considered "publishable" and is often considered to be a "dead end". If a practice, technique, interpretation, theory, or tactic, is working well within the coherent social agency of the labour process, then it will be used whilst the experimenters deal with more interesting or problematical matters. It is this pragmatic acceptance of practices within coherent social agency that establishes stable technological objects according to their instrumentality. The stable technological object is made pragmatically meaningful through training and demonstration in such a way as to "explain" the meaning of the practices by showing its intended effect in the context of the work. The purposes of practices are given a pragmatic basis of meaning in terms of their usefulness, which the student learns through mimetic attendance towards the uses, the intentions, and the expectations, implicit within the labour process. However, a research laboratory is not comprised of homogeneous individuals and does not constitute a simple social agency.

Physicists, like people from all walks of life, are idiosyncratic and have divergent interests, and constitute complicated social agents in their own right, which may, or may not, cohere with the aims, expectations, and practices of other social agents. Within any laboratory there is a diversity of opinion about "how to do physics", "how to solve this particular problem", or even if "there is a problem here". Of course, the practices of any working experimenter must remain within the shared template of the methodology if they are to be performing the same experiment and communicating their experiences in commensurable terms. From within the boundaries and constraints of the shared methodology the experimenter has considerable space for negotiation and choice about her/his response to perceptions of competence, credibility, and the potential outcome of interpersonal interactions. Practitioners may well share a commitment to make the experiments they perform work well, and to publish excellent results, but they may not necessarily agree about what constitutes "working well" and "excellent results". The diverse opinions on interpretations, visualizations, or appropriate techniques, can be sufficiently incoherent as to leave a great deal of latitude regarding what constitutes a "skilled judgement". In the process of experimental work there is a dynamic social process of bringing-together diverse social agents engaged in the negotiation of "how to proceed". This dynamic social process is directed to transform a collection of diverse social agents into a unified convergent social agency in interaction with other laboratories, conferences, and funding bodies, etc. Many of the choices required for the construction of "skilled judgements" are not always made on overtly "scientific" criteria. Economic and political factors are involved in choice of projects in order to gain prestige for the laboratory, attract funding, attract media attention, undermine a rival laboratory, etc. Choices are bounded, context dependent, and pluralistic selections because they are made in contexts of resource relationships, interests, and social connections that transcend the laboratory, based on opportunism and estimations of risk within a wider social context, and are not made solely in terms of internal logic, or "scientific" rationality. As a social agent, that being a scientific negotiator involves being a negotiator in a wider context than the laboratory.

Through the above social processes, the student, by acquiring competency, is orientated within the developing stable labour processes of the group in such a way as to become a negotiator within the group in terms of a praktognosic exchange of "skilled judgements". A stable labour process constitutes simultaneously both a stable social order and a stable technological order. It is a socio-technical organization of technological objects, bodily motility, and technique in accordance with the demands of research. In the context of production, "order" is determined pragmatically in terms of "stability" and "reproduction". Any change of the labour processes will occur in response to "disorder" and "instability" and will become established if it is deemed to promote "order" and "stability". In this way the technological framework develops in accordance with ongoing stabilization of labour processes in response to periods of instability and disorder. The established convention of the technological framework is periodically open to re-evaluation and negotiation as periods of incoherence, disorder, and instability regarding "the best way to proceed" are encountered. These periods of incoherence manifest themselves as new problems and the awareness of a lack of "skilled judgements" regarding the solution to these new problems. These new problems arise because in innovative and complicated work, such as experimental physics, the socio-technological order is *underdetermined* because technologies are used for purposes other than the use for which they were developed and, consequently, their use, in terms of what can be done with them, remains to be constructed through trial and error judgements from within a template of possibilities. The interaction between changes in the technological configurations of the apparatus and its responses to those changes are also underdetermined (otherwise there would be nothing to experiment upon) and opens a space for choice in the selections and directions available to physicists. It is this space for choice that allows considerable creativity and free-play in scientific work in which incoherence between social agents is possible due to the availability of a plurality of divergent selections and directions. This space allows an experiment to begin and is removed once a coherent consensus is made regarding the achievement of stability within the labour processes and the technological configuration of the laboratory. It is impossible to determine the content of an ordered labour process from the onset because it is developed experimentally through periods of disorder and ordering. In other words, stable productivity is incomplete, experimental, and no one is able to determine, in advance, exactly what form it will finally take when it is complete. The working physicists may well have expectations, as to what the final form might be, but these expectations are re-evaluated and transformed during the process of ordering the social and technological dimensions of the work. Provided that they remain within the template of the methodology of their institutionalised field of research, they have considerable free-play in the construction of their plan of action. Furthermore, the technological objects used in experimental work were constructed in heterogeneous contexts. Each object is teleologically posited as a transformative power when it is stabilised as standing-reserve. An innovative process is one of converging heterogeneous objects and integrating these diverse centres of transformative powers into a single convergent, stable, and coherent centre of transformative power in the novel context. This creates a hybrid novel prototype, available as a technological object for future experiments. It is also how physics progresses and specialization occurs. It is heterogeneous, fragile, and fragmentary. It is experimental.

Experimentation is a highly developed form of social activity that, for all but the most trivial experiments, requires co-operation, resources, and complex machinery. Each physicist, when positing the value of the experiment, is positing purposes and goals to other people that the experiment has, at least in part, the potential to satisfy. Thus teleological positings for any experiment is two-fold: the experiment must be presented as having an external value (i.e. value for economic, military, or political purposes) and as having an internal value for the ongoing activities of research (i.e. value for scientific purposes). It has an "external" value in which the co-operation of non-physicists is required to provide the resources and facilities for the construction of a space in which an experiment can occur. It also has an "internal" value in which the co-operation between physicists is required for the successful design, construction, operation, inscription, and interpretation of the experiment. Each experiment must be set-up as instrumental for the satisfaction of purposes and values, which are built into the set-up (unless the physicists involved are being deceptive) and bounds the estimation of what constitutes technically rational skilled judgements regarding the stability of the labour process. Physicists, in order to persuade themselves and others of the experiment's potential, must do so in relation to a background of conventions, expectations, estimations, and acceptance of what is possible, probable, or even necessary. Thus every experiment is set-up in relation to a paradigmatic background, against which it is posited as an exemplar, within the "internal" trajectory of ongoing research and the "external" trajectories of the wider world's desires for further technological innovation and power. From within these trajectories particular exemplary experiments are publicly presented as necessary (or crucial) for the further progression of those trajectories. In this sense, experiments are suggested by paradigmatic backgrounds of ongoing scientific research and technological innovation. The physicists are challenged by these suggestions to set-up and perform the experiments that publicly emerge as necessary, crucial, probable, or even merely possible. From within complex experiments a further series of related experiments emerge as necessary, crucial, probable, and possible, and the participants will be challenged to construct and perform them next. It is in this sense that an historical understanding of the ground plan of the set-up and trajectory of experiments is *necessary* for us to have an understanding of contemporary experimental physics.

The trajectories of ongoing research and technological innovation are emergent as teleological positings from the technological framework. The highly complex machines and infrastructure required by modern physics require division of labour in order to function successfully as research. The individual participants are defined in terms of their roles (i.e. physicists, engineers, technicians, mathematicians, students, theoreticians, and computer scientists) as they are ordered according to their techniques, under the challenging and destining of *Ge-stell*. Each one of these techniques is defined according to the postulation of a purpose within the complex of purposes ordered towards the purpose of the whole. Each participant attempts to elicit performances from their part of the machine interface and from other human participants. The object of their labour is not just the performances of the machine but is also the performances of the whole, interconnected complex of participants upon which the performativity of the machine depends. Thus the enframing means to achieve machine performativity are orientated towards both the machine and the human participants. Machine performativity emerges from the *Ge-stell* of labour upon materials, inscriptions, and other people, and is not a simple test of theory. The purposes and goals postulated in such complex experiments constitute technological objects in themselves and are not simply rational orientations towards theoretical derivations. The teleological positings of labour are the challenges of *Ge-stell* and where "theory testing" plays a role in an experiment it is as a challenge within the complex of interconnected challenges set-up by *Ge-stell*. If complex experiments can be said to be a test then it is a test of the socio-technical agency that is gathered and challenged to construct and perform the experiment. It does not immediately follow from the success of agency that any theory utilised in the experiment was correct (or even approximately correct). All that immediately follows is that yet another challenge has been undertaken and completed.

The stabilization of any experiment results in the concrete realization of a techno-phenomenon that is produced through the contingent activities and choices made during the efforts to stabilise those practices. It does not follow that techno-phenomena realised in this way were waiting to be discovered, or were actualised in accordance with natural law. What is discovered is how to stabilise those practices. In my view, J.J. Thompson did not discover the electron as something waiting to be discovered but, rather, discovered how to make the electron a stable part of the ongoing extension of the family of electromagnetic machines. Given that the electron only exercises its powers within the contexts of this machine-family, then it is irrelevant whether it is a "real and out there" fundamental corpuscle of matter because "what an electron is" emerges as an exoreference within the technological framework of the ongoing researches of experimental physics. Its reality is inextricably bound up with the destining of these researches and it is "that electron" which is the object of scientific discourse. The electron is a technological object and its reality should not be divorced from the socio-technical processes in which it is stabilised and utilised. The technique of spraying electrons, for example, is available as standing-reserve for challenges by *Ge-stell* to discover free-quarks and its disclosure is identical with the responses of machine performances to human interventions. On this account, the electron of scientific discourse does not need to have any scientific reality outside of the technological framework of research because if it were to have ontological independence from the technological framework then these facets of its being will not be utilisable as standing-reserve and, consequently, be inaccessible to scientific research. It would be outside the technological framework. The labour processes involved in the stabilization of technological objects requires bringing together and ordering diverse and heterogeneous technological objects across the borders between machine-families. These objects can be skills, practices, machines, exoreferences, models, metaphors, representations, mechanisms, tools, materials, techniques, or inscriptions. The innovative processes of stabilization involve the convergence of divergent and incoherent centres of transformative power into a single convergent and coherent centre of transformative power. This is a process of producing a stable technological object from many technological objects within the complex manifestations of a historical labour process. These emerge from a struggle with contradictions, incoherence, divergence, heterogeneity, and plurality that are implicit in, and emergent from, the extension and synthesis of distinct machine-kinds into a novel machine-kind. Technological objects made for the purpose of investigating Nature are used to produce intelligible models of natural phenomena, and the ongoing development and extension of physics towards its own perfection as a *techne*. Implicit in the concept of experimental investigation is the conception of the reality of those technological objects in terms of their instrumentality in material practice. Thus Hacking was a realist about electrons when physicists claimed to spray them on mobidium spheres as part of the process of investigating the existence or inexistence of free quarks. However, the ontological status of postulated entities does not only depend on their instrumentality. Otherwise Hacking would be a realist about G-7 chords because they can be played on a number of different stringed instruments. Both electrons and G-7 chords are transfactual indices that arise through particular modes of disclosure and are implemented, as principles of organization and cognition, in the interaction between human agents and machines (or instruments). However, electrons are not usually given an equivalent ontological status as G-7 chords because the latter index is not usually taken to have a correspondent referent in the natural world. The presumption of mechanical realism allows the production of technological objects such as electrons to be presented as a product of a fundamental relation between human intervention and natural processes that is a consequent of invoking natural laws by designing, constructing, inscribing, and interpreting machine performances. Once this is stabilised it is presented as a defined and self-contained technological object utilisable for the exploration of a "deeper and more unified strata of reality" and the innovation of novel machine-families is taken by scientific realists such as Bhaskar to be a "deeper" exploration, disclosure, and discovery of natural mechanisms, causes, and laws. Thus the extension and synthesis of a compass (a magnetic machine) and a chemical battery and wire combine (an electrical machine) into Oersted's apparatus could be represented and utilised as a technological object for the disclosure of the deeper and more unified strata of "electromagnetic forces". In Lukács' terms, this act of alienation was one that involves the abstraction and reification of labour.[14]

Technology, truth, and experimental physics

Modern experimental physics is governed by craftsmanship and modern technology whilst it is bound-up with *Ge-stell*, *poiesis*, and *techne*. *Ge-stell* operates upon experimental physics, whilst it is internally directed towards the acquisition of its own *techne*, as a poietic mode of disclosing the alethic possibilities of reality as standing-reserve available for future experimentation and technological innovation. The convergence operation of *Ge-stell* upon machines provides technological objects with transfactuality by gathering and ordering them into strata and clusters of machine-kinds. This innovative historical extension of technological objects into novel machine-families provides experimental physics with stratification. These provide both of Bhaskar's ontological dimensions. The transfactuality and stratification of the technological objects utilised in experiments provides physicists with senses of ontological "depth" and discovery through the process of making innovation intelligible in terms of models. The labour processes involved in the construction of stability and the convergence of practices towards objectivity, in the artificial contexts of experimentation within closed systems, are nothing more nor less than the construction of repetitions that can be described technographically and presented as empirical regularities. The physicist is discovering how to make the act of making intelligible in terms of the interconnected strata of machines and transfactual technological objects upon which the experiment has set upon. Once mechanical realism has been presumed then techneic causal accounts, operating with technographic functives and visual representations can be metaphorically used as mechanical models of the natural phenomena under investigation. This is essentially a process of "reverse engineering" in which the physicists construct a mechanistic model of the machine performances in order to imagine the "natural machine", in operation behind appearances, which generates the artificial machine performance. The physicists then compare the expected performance with the actual performance. By invoking the natural economy of mechanisms, when similarity increases, the physicists become increasingly confident that the precepts of mechanical realism underwrite the removal of technological processes from the final accounts. This permits machine performance to be treated as a transparent mode of disclosure, and, the *techneic* account of the causal series stabilised during the production of empirical regularities can then be presented as the ontological law that was disclosed by the experiment. This metaphysical foundation allows *techne* to be presented as *episteme*, provides experimental physics with an ontological dimension, and the achievement of stability facilitates the abstraction and reification of *techne* as natural law. Experimental physics is a labour process that uses models and metaphors in order to produce intelligible accounts of natural phenomena in terms of mechanisms and laws. It uses technological objects to produce further technological objects to satisfy the "internal" challenges of scientific research and the "external" challenges of the wider world of economic, political, and military ambitions. The craft practices and technological trajectories of experimental physics are directed towards the production of intelligible mechanistic models and prototypes, and, that the pursuit of modern experimental physics is a mode of both *Ge-stell* and *poiesis* that participates in "world-making".

The difference between craft practices and modern industrial technology is that the latter drives towards the creation of new instruments, in response to new needs, whilst the former continued by extending, refining, and perfecting the same means to achieve the same ends. Modern experimental physics operates upon the boundary between craft practices and industrial technology. It is both radically driven towards novelty and is conservatively attendant to its own self-perfection and refinement of established means. It is challenged to perpetually innovate new machines and instruments, whilst extending, refining, and perfecting itself as an alethic disclosure of itself as a "bringing-forth" of the truth. It is destined to order itself as standing-reserve for future work and brings beings into the world for their own sake. For example, the electromagnet started as an underdetermined object for experimentation in the work of the eighteenth century physicists but, in contemporary physics, it has been stabilised as a technological object to be used for determined and repeatable purposes. It is a component in the cooling process in ultra-low temperature physics experiments performed on the superfluid properties of He-3 and it is a component in the focussing of electronic and positronic beams in the CERN experiments on the properties of fundamental particles. The performances of electromagnets have transformed from being the end of research to being a means for research into the performances of other objects. The connection between *poiesis* and *Ge-stell* is premised upon the conception of change as being the product of the exercising of a "natural mechanism". The "electromagnetic field" is a complex technological object (a composite of other technological objects) available for future work as a tool and is also a unified object for reflection in its own right. This two-fold character of the objects of physics is a manifestation of the two-fold character of physics as it operates across both spheres of crafts and industrial technology as a bridge between the two. Modern experimental physics has its origins in the craft practices of the sixteenth century and provided the conditions for the industrial machinery of the nineteenth century. The mechanical realist precepts made this bridging possible.

When Ellul wrote "the search [for efficiency] is no longer personal, experimental, workmanlike; it is abstract, mathematical, and industrial", he did so without a close inspection of the technical practices of experimental physics.[15] In experimental physics the search for efficiency is personal, experimental, workmanlike, and, it is abstract, mathematical, and industrial. The "natural mechanisms" disclosed by experimental work are simultaneously treated as truths and potential instruments for the ongoing trajectories of research. If spontaneity and chance are eliminated by the technical imperative (as Ellul argued that they are) then experimental physics cannot be circumscribed by technique (as Ellul defined it). Without spontaneity and chance there is no space for discovery and innovation. It is for this reason that experimental physics must remain on the border between craft practices and modern technology. Furthermore, the performativity of the objects of experiments cannot be determined in advance (otherwise there would not be an experiment at all). The underdetermined character of the technological objects studied in experimental physics, as objects of research, lends research a need for the innovative attentiveness of the "know how" and skills of the experimenters as craft practitioners. It is this need for innovative attentiveness and craft practices that makes experimental physics an art. This is a necessary condition for experimentation because, as an art, experimental physics allows a dialectical variability, in the construction and performance of an experiment, which is a condition for being an experiment at all. It is the underdetermined character of the objects of research that provides physics with an experimental character because the objects of experiment need to be stabilised, in terms of mechanical and repeatable performances, before they can be assembled and mechanically integrated into industrial production. Once this has been achieved then the performative character of the objects has been determined and, consequently, is no longer of any experimental interest. Thus the productive aspect of physics is the disclosure of mechanisms as standing-reserve for *Ge-stell*, whilst the poetical aspect involves situating these disclosed mechanisms within a world-picture. The *Ge-stell* aspect of physics is the challenging of *poiesis* to bring-forth mechanisms for the sake of disclosing them whilst is disclosed as truth by successfully implementing them in productive practices. Experimental physics is artistically and instrumentally situated between reiterated feedback loops between the production of intelligible information as an internal good and as a technological object for future implementation.

If we treat technology as the making and using of tools, techniques, procedures, materials, resources, skills, and machines, as means to ends, as embodied in our everyday practices, then it is largely an unreflective activity directed towards unreflected-upon ends, which produces and manipulates unreflected-upon objects. Artifice is the skill of how and when to use specific technologies in order to manipulate things in the world to achieve specific desired goods (or ends). We largely take it for granted in the modern world. By embodying artifice, human beings are situated and immersed within a technological frame-work that extends beyond the individual human body. Technologies are socially organised structures that, through the embodiment of artifice, are integrated into the agency of the individual human body and empower it. As Stiegler put it,

> For to make use of his hands, no longer to have paws, is to manipu-
> late – and what hands manipulate are tools and instruments. The
> hand is that hand only insofar as it allows access to art, to artifice,
> and to *techne*.[16]

Artifice involves a spatial and temporal ordering of productive practices, in order to effect change, transform things into other things, and brings things into the world.[17] Artifice empowers and directs human bodies to effect change. It is an ordering structure that directs the practices through which things are brought together, arranged in temporal and spatial sequences, and transformed into other things. Artifice is effective through being embodied in human activities involving materials and inscriptions. In order to become technological, a human being must embody the material and inscriptive practices that artifice imposes upon the organisation of her/his motility and intentionality. The would-be skilled practitioner must learn how to use specific tools to perform specific activities, upon specific materials, to transform those materials into specific products. The process of embodiment of the artifice of practices upon materials is an embodiment and exercise of the discipline/power of the technological framework that generates pro-ductive human agency *qua* technological agency upon specific things in the world. This is enabled towards a horizon of specific ends or goods. Embodied artifice upon materials towards a horizon of specific ends is a process of *poiesis*: bringing entities forth into the world. It is this aspect of crafts that for Aristotle, and Heidegger, related *techne* to *phusis* and distinguished each from the other. *Techne* "brought forth" the object through the agency of the craftsman upon the materials. *Phusis* "brought forth" itself without the aid of the craftsman. *Phusis* was bringing forth in the highest sense, for both Aristotle and Heidegger, and it is this relation that situates *techne* within the history of metaphysics.

Artifice extends the agency of human body. This extensional relation between tools and the human body has existential import upon human agency in terms of power, identity, and the horizon of our alethic possibilities. The discipline/power of artifice is exercised through learnt modes of labour directed towards efficiency, productivity, and perfection. Artifice extends human manipulative power and the field of material practices for achieving specified ends and, by constraining and empowering human agency, artifice channels and extends human agency towards an otherwise unobtainable horizon of possibilities. Artifacts, such as computers, measuring instruments, and maps, organise and order technographe to extend the human capacity to manipulate, record, and visualise abstractions. Instruments, such as thermometers and weighing scales, exoreference changes in the world in terms of abstract quantities. These devices are effective in virtue of being embedded in a set of interpretive and inscriptive practices that enhance and channel the human cognitive imagination into extended capacities for intellectual agency at the expense of absolute freedom of thought. The human agent can exert considerable creativity and free play when using these technologies only to the extent that s/he acquires the appropriate artifice for achieving specified ends (an otherwise unobtainable horizon of projected alethic modalities) and remains within the technological framework of the conduits of disciplined power. By providing an otherwise ineffective and undisciplined body with agency, artifice permits the existence of specific intentions and provides us with both means and ends. Artifice shapes intentionality and agency by shaping the horizon of possibilities available to us, as well as the ways of reaching that horizon.

Human life entails an interactive relationship between intentionality and artifice, and, consequently, "human nature" is not given by "Nature" but is created by disciplined and empowered agencies projected beyond "organic necessities" in terms of self-interpretive and self-creative destining that starts from a challenge and ends with its material realization as human beings utilise techniques within a horizon of intentionality. The interactive relationship between intentionality and artifice, in order to realise projects, and consequently "human nature", is itself technically mediated through the template of how to proceed and the challenges that are received alongside the power of technological agency. The centrality of technique to human technological existence and "human nature" provides existential and aesthetic import of technology to human life and character. At least in part, we are defined by what we can do. However, there is not a singular set of techniques for each and every project and, even as technological agents, we are open to alternatives. For some projects there are (currently, at least) no available techniques at all (e.g. interstellar travel, world peace, immortality, or time travel) and consequently they remain in the human imagination, fears, dreams, and desires. They are so remote from our bounded technical rationality that they are barely challenges. For other projects there are a plurality of techniques (e.g. fossil fuel power stations, nuclear power stations, building wind power turbines, tidal power ducts, and solar power technologies are distinct techniques for electricity generation) and our bounded technical rationality involves the problem of how to choose between them. We should not be radically individualistic and locate the origin of challenges and the destining of their satisfaction in the imagination of the human-subject. This would ignore the social origins of human aims and ends. It would ignore the teleological character of human life and labour. Artifice, by shaping both our horizon of possibilities and the means by which we reach it, provides both intentionality and techniques. Human agency, to the extent it is defined in terms of what human beings can do (as well as by what we can not do), is constructed through artifice and is not simply created by human beings. If we take technology to be the means by which we move from the challenge to realization then technique is itself bound up with the teleology of artifice.

Can we moderns make sense of Aristotle's ancient claim that *techne* "resides in the soul of the craftsman" and that the products of *poiesis* find their origin in the producer? As a technological framework of practices and materials, artifice can only be said to be the property of a human agent in virtue of its successful embodiment. *Techne* would reside in the soul of the craftsman *qua* technological agent. If the soul of the eye is "to see" then the soul of the craftsman is "to craft". The soul would be the destining of the craftsman *qua* craftsman. This soul is itself born through the embodiment of artifice as a human body is transformed into a craftsman through such an embodiment. The products of *poiesis* find their origin in "the producer" in the sense that a human body is transformed into "the producer" through the successful embodiment of the disciple/power of ordered practice. Artifice is concealed during its embodiment in practice. Mastery occurs through the successful embodiment of practices as habitual practices conducted with confidence and productivity. The acquisition of technical expertise is not a mastery over artifice itself but, rather, a process of publicly becoming one of artifice's competent servants. The skilled craftsman is an exemplar of artifice rather than a master of it. The apprentice imitates the practices of the craftsman and, by doing so, participates in the perpetuation and dissemination of artifice. The craftsman, as an exemplar, is the focus of the apprentice's attention as source of artifice and, consequently, the illusion of mastery is propagated. Furthermore, as the horizon of possibilities surrounds human life and artifice is taken to be the property of human masters, the intentionality made possible through artifice is taken to be simple human ends. Artifice is concealed as human means to human ends. It is in this sense that *techne* could be said to reside in the soul of the craftsman. "Human nature", in part, is created by the power/discipline of artifice in the construction of human agency and the destining of the projected horizon of alethic modalities.[18] It is for this reason that we should not locate control and intentionality within the human-subject. It is only once the embodiment of artifice has become transparent during the construction of the human agent *qua* technological agent, that control and intentionality can be taken to be the property of the human agent. The centre of control and intentionality lies within the technological framework of a labour process that is accessible through the embodiment of artifice itself. It operates between the human body and the objects to be transformed through discipline/power and embodiment relations of artifice. *Poiesis* emerges as a labour process of feedback relations, mediated by artifice, between the human body and the objects to be transformed in accordance with an anticipation of *techne*. It is primarily a productive process teleologically posited as the means to manufacture specific products through feedback adjustments occurring between the practices, technologies, and the materials to be transformed. This kind of labour process is an economic process in which artifice and its horizon of specific goods are primary social relationships of exchange value.

Invention proceeds from a cultural technical background of artifices, technologies, tactics, and challenges, towards specific end products. Artifices, technologies, and tactics are selected from this cultural background with the aim of constructing an artifice, technology, and *techne*, as not only a means to produce the desired product, but as a means to produce a means of production. This constructive process involves the convergence of heterogeneous technological objects from strata of machine-families and integrating them into a unified technological object in accordance with a specific goal. These diverse artifices, technologies, and challenges, are brought-together, ordered, and integrated towards the projected asymptotic horizon of *techne* with the aim of achieving that ideal knowledge. Diverse artifices, technologies, and technological objects each constitute a centre of transformative discipline/power in their own right. The process of integration requires bringing-together these diverse centres of transformative discipline/power into a coherent single centre of transformative discipline/power. Constructing a new process of destining is itself the object of invention. The construction of destining, the conclusion of the ordering process, is itself the construction of the desired artifice. This process is an *undetermined* process. We cannot know, from the onset, whether or not an artifice, and its technologies, can be produced to produce the desired outcome. Whether a time-travel machine, an anagathic anti-ageing pill, or a cure for cancer can be invented is undetermined. This remains undetermined even when the desired product is produced because the "causality" of any technology remains complex within the wider world. For example, the destining, challenges, and transformative powers of even burning coal have yet to be determined because we have yet to categorise and provide causal accounts for all of the consequences of this material practice.

Novel experiments must occur without *techne* or else there would not be an experiment. If experimentation is to be understood as a form of artifice then we need to define a different kind of productive process. This is the innovative process in which artifice is itself formed, transformed, and brought-forth as the means to extend and complete itself as an art. Innovative processes are more sophisticated than economic processes. Processes of this kind are both the subject and the object of themselves. An innovative process is fed-back onto itself as both its own means and its own end. In these processes the development of the process is itself the product of itself. It is a non-linear process that turns upon its own possibilities as its own resources and final end product. It involves the feedback of the uncontrolled control of control back into the process as control-as-information for future control and the relinquishing of control. In experimental physics, as in music or any art using instruments, this involves mapping out the contours of human interventions and machine performativity. It is distinct from invention, which is the economic process by which a novel form of artifice is itself produced in order to achieve a specific challenge and what is lacking is the artifice to achieve it. In contrast to invention, the "trial and error" processes of innovation are an extension of the technological background as an art extended towards its own perfection by providing a complete set of alethic modalities for their own sake. The ideal generalization of this process into an abstract and communicable form of knowledge is the construction of a *techne* that could be disseminated as instructions of how to repeat the process of production as a *mathesis* of its own alethic truth.

Experimental physics is only merely inventive when the aim of the experiment is to construct an apparatus to perform a specific task, measurement, or manipulation. Thompson was inventive when he constructed a means to measure the charge to mass ratio of a cathode ray by devising the means to manipulate and inscribe a cathode ray in a cathode ray tube. The destining of this experiment towards the ideal *techne* of the first causes involved in repeating this experiment brought the Thompson experiment into the public realm as a means to disclose a deeper fundamental level of reality. However, the claim to possess *techne* requires that the production process is completely causally understood, but the claim for the truth of this kind of causal account is deferred until it is successfully implemented in future innovative practices. This will involve further productive practices and an anticipation of a deeper causal account that, in turn, will be deferred until it is successfully implemented in future innovative practices. The final end of this process of deferment and innovation could not cease until further innovation was impossible and it is only at that point could *techne* be graspable. Innovation occurs when the productive possibilities of artifice are explored in the absence of a specific end product, for its own sake, as an alethic process of mapping out the horizon of productive possibilities and the routes that a particular artifice takes to reach them. When a musician explores the productive possibilities of her/his instrument, without aiming to compose a specific musical piece, s/he is engaged in a process of innovation. The process of innovation is the process of exploring and determining the productive possibilities of artifice. The international community of physicists and engineers at CERN are in the process of exploring the productive possibilities of the LEP ring and LHC machines by innovating them as a means of exploring the alethic truth of these machines. These physicists are exploring the productive possibilities of these machines in order to understand how they work in terms of alethic causal accounts. These projects are challenged towards the innovation of all the possibilities until the *techne* of the whole productive process has been achieved. The physicist learns that when s/he performs a particular intervention then the machine performs in a particular way. Thus the physicist can make a mapping between a range of particular interventions and a range of particular machine performances and inscribing these contours with technographe and producing mathematical inscriptions of novel transformative powers.

Each machine is an integrated composite of components (which are machines themselves) innovated to capture the *mathesis* of its functionality. It is the sum total of an integrated nexus of distinct centres of transformative powers coherently converged into a single centre of transformative power and a source of truth. Associated with each machine are collections of artifices to design, construct, operate, maintain, repair, and interpret the machine performativity related by a collection of fragmentary specialised templates to build the machine and its components. These collections and relations, for even moderately complicated machines, are distributed throughout the whole teleological organization of the labour process in which the machine can be brought into existence and integrated as a functioning entity and disclosure of truth. This will involve the division of the embodiment of these collections among many human agents that are themselves transformed into functioning technological objects within the whole technological framework. In experimental physics, as well as engineering and other technosciences, the collections of technologies, specialised tools, mathematical techniques, functives, and technographe, associated with machine-kinds, phenomenologically constitute autonomous modes of agency in their own right. Machine autonomy is experienced as an integration of transformative powers together to produce a unified transformative power emergent through performance. Experimental physics, alongside all arts, innovatively explores reality through disclosing the productive possibilities available to human agents by experiencing the autonomy of their performances. It is in this context that experimental physics could be said to be a metaphysical performance art in the highest sense. However, the reality disclosed through this performance art could not be said to be independent of the modes of disclosure innovated to disclose it. It is not objective in the usual sense of the term. Nor could it said to be simply subjective. The metaphysical precepts of mechanical realism underwrite the experimental disclosure of the truth of reality that is interactively made through technology and tested in terms of its availability for future use. The ontology of machine-families provides physics with a technical-material infrastructure that is its proven in terms of its standing-reserve as a technical background and concrete reality. The specification of "a universe" as the object of technoscientific activity is the metaphorical substitution of the techneic knowledge of machine performativity for the epistemic knowledge of the natural world. The "know why" questioning of scientific questioning is characterised by the question of why something works and, consequently, *episteme* is reduced to the "know how" causal account of an ideal and unreachable *techne*. As both of these kinds of knowledge are ideals then the whole final truth of physics is anticipated through the ambiguous metaphorical interchanges between two imaginary ideals.

6
What Enables Us to Build Machines?

The innovation of electricity, nuclear power, aeroplanes, rocketry, plastics, computers, etc., has changed the world. However, do scientists understand the powers that they release into the world? What enables us to build machines? This question is a central question for any critique of realist and positivist positions because both will claim that acting upon materials in accordance with the "natural law" will enable us to build machines. The irony is that both positions neglect the productive role of technology in experimental physics, and yet their arguments for the success of physics are premised upon its technological success. My argument against both realists and positivists is that we cannot find any rational or empirical grounds for deciding between physics-as-discovery and physics-as-production once we address the technological character of experimentation. What physics does is produce the process of learning how to create itself as the means to produce its discoveries. As I argued in chapters three and four, the metaphysical project of experimental physics involved understanding natural processes in technological terms from the onset. This has been evident throughout all of experimental physics, including the development of mechanics, optics, thermodynamics, electromagnetism, and quantum physics. The history of physics is itself comprised of innovating, bringing-together, developing, and explaining a series of heterogeneous machine prototypes. These machines are not reducible to the others, yet the dynamic scientific project of explaining the possibility of their innovation interrelates and stratifies them within the ontology of physics. The mechanical, thermodynamic, optical, electromagnetic, quantum machine-kinds are all assembled and interrelated within the context of ongoing exoframed labour processes directed towards the acquisition of the *techne* of Nature. *Techne* promises to provide the complete alethic knowledge of how to bring things into existence and mechanical realism underwrites the epistemic idealization of that unreachable asymptote. The existence of a *techneic* law was teleologically posited as something to work towards, in accordance with the ground plan of constructing a machine as a demonstration and disclosure of it, as physicists "test" and "refine" models by implementing them in future technological innovation. This perpetually ongoing labour process indefinitely defers the determination of the truth-status of any theoretical representations, whilst instrumentally using them to make interventions in ongoing experimental work. As was argued in chapters four and five, the ontological stratification of experimental physics is delimited by the machine-families developed by that science. All machine-kinds are collections of prototypes within machine-families that are linked by shared technological objects, techniques, materials, and general models. This shared stock between machine-kinds gives machine performances their transfactuality. Bhaskar was right to criticise the positivistic view, which holds that constant conjunctions of the form "whenever X then Y" are the epistemological limits of scientific investigation, as a mistake that has been disastrous for our understanding of science.[1] Of course one must be able to state that "whenever X then Y", as a starting point in the description and design of machine performances, but this is insufficient. One must also state the mechanism(s) that occur between X and Y before one can hope to propose an understanding of how machines work. I agree with Bhaskar that limiting an account of the epistemological goal of experimental sciences to the identification of constant conjunctions is a mistake, but he missed a crucial point about the ontological relation between such sciences and the machines that they use. The ontological relations of experimental physics are identical with the metaphysical interpretation of the interactions between human interventions and machine performances. Experimental physics can only extend its ontology – engage in a process of discovery – by producing new machines. Specializations inform each other and are commensurable in so far as they use each other's machines and techniques. The progress towards complete alethic knowledge is one of increasingly enframing the world within the metaphysical and technological framework of machine performances. It is the world-picture, abstracted from this whole network of metaphysically conceptualised machines, which is ontologically posited as "the known universe". Positivism and empiricism may well be disastrous to our understanding of the sciences but, in my view, mechanical realism may well prove disastrous for our understanding of Nature.

Experimental physics is a produced means of production that is both "tested" and made meaningful within processes of ongoing technological innovation. The objects of scientific thought have phenomenological "independent existence and activity" because they are not entirely mental entities and are not entirely controlled by human intentionality. The objects of scientific thought elicit the experience of phenomenological autonomy because they are technological objects emergent and operational within a technological framework that transcends, unifies, and structures the particularities of practice and experience. A technological object cannot exist without the conditions of its production, but its trajectories transcend human control and expectations. Such objects are not simply "man-made" but, despite their phenomenological reality, they do not independently exist in the realist sense either. In order to understand technological objects we need to move beyond linear dualities and address the non-linearity of human-machine relationships. In order to understand the historical development of material practices at work in the creation of new scientific theories and technologies, as the product of creative labour processes, we need to address the ways that technological objects are fed-back into the processes of experimentation and the novel extension of the variety of available technological objects. Each theory is metaphorically postulated as representative of a natural process by social imagination, whereas it is technically produced by the ongoing innovation of a machine-family and tested by its instrumentality in the extension of that machine-family. Once human imagination, ambition, and metaphysical justification have been culturally embodied in ongoing material practices directed towards the acquisition of complete alethic knowledge, then its abstractions can be poetically described by the metaphor of "Natural Law" and the "book of Nature" can be read from sheets of mathematics and diagrams.

How can we explain the technological success and the explanatory success of physics without conceding scientific realism? The theory of Quantum Electrodynamics (QED) may well have unprecedented predictive success, but what does it predict? Its accurate predictions of the magnitude of magnetic dipole moments, for example, are the anticipation of how a particular kind of machine should perform. Scientific knowledge is neither "read straight from the natural world" nor "out of the human mind". This is a false dichotomy. Scientific knowledge is the postulated goal of the processes of modelling the contours of human interventions and machine performances. This social imagination, ambition, and justification, within the context of technological innovation, allows machines to act as interface between human beings and postulated natural mechanisms and laws. The machine performances are complex, neither entirely dependent upon nor entirely independent from human agency, because the centre of control of the labour process lies neither in a "material world" nor in a "human world" but rather occurs during the processes of bringing together diverse agents and attempting to converge and integrate them into a single unified centre of control. It is this project of integration and unification that constitutes the context of technological innovation. Human interventions and machine performativity, mediated by models and techniques, are gathered-together, shaped, and ordered in accordance with imperatives and challenges of *Ge-stell*. Technological objects have degrees of autonomy once they are embodied in diverse productive practices directed to unify and communicate new experiences. The uncontrollability of technological objects, which is characteristic of their intransitivity, is due to the complexity, incompleteness, diversity, and incoherence of the whole experimental process of innovation. This is a consequence of the heterogeneity of the technological objects used in the technological framework of experimentation. The autonomy of experience within a technological framework is itself the product of the metaphysical art of experimental physics, perpetually directed towards the unification, completion, and perfection of itself.

Whence the resistance?

In *The Mangle of Practice*, Andrew Pickering presented an interpretation in which the results of experimental physics emerge from a dialectical relationship between "human agency" and "material agency" that occurs on the interface of machine performativity.[2] His interpretation of modern experimental physics is that it is a performative and productive social labour process comprised of material practices that are dialectically transformed during real-time accommodations to resistances. He argued that "material resistances" to human intentions, and the accommodations made to plans and intentions in response to those resistances, demonstrate that materials are agents. Using the example of Glaser's attempt to "build a bubble-chamber", Pickering pointed out that Glaser had to find many different solutions to "the triggering problem", during the course of developing a working bubble-chamber that could detect "cosmic rays". However, each proposed solution failed, one after the other, despite Glaser's expectations of success with each new solution. Pickering asked the following questions: if each of these "solutions" were socially constructed as "expected successes", and "the detection of cosmic rays in bubble-chambers" is also socially constructed, then, why should we see this sequence of failures? Where was the social causal factor here? Who was constructing the failures? How can the strong programme in the sociology of science explain the unexpected failures in the history of physics? Why does failure occur when the prevalent social consensus and dominant authority expects success? Pickering argued that the strong programme does not provide an intelligible account of scientific practice because it neglects "material agency", which, in interaction with human agency, transforms scientific practice. He argued that the interests and identities of scientific agents are at stake within scientific practices rather than causing and explaining the extension of scientific culture. Hence, Pickering wrote that

> [i]t is clear that Glaser had no way of knowing in advance that most of his attempts to go beyond the cloud chamber would fail but that his prototype bubble-chamber would succeed, or that most of his attempts to turn the bubble-chamber into a practical experimental device would fail but that the quenched xenon chamber would succeed. In fact, nothing identifiably present when he embarked on these passages of practice determined the future evolution of the material configuration of the chamber and its powers. Glaser had to find out, in the real time of practice, what the contours of material agency might be.[3]

Pickering defined material agency as "simply the sense that Glaser's detectors *did* things – boiling explosively or along the lines of tracks or whatever – and that these doings were importantly separate from Glaser."[4] His point was that the state of affairs, which arose in the performance of the machine through Glaser's relation with the bubble chamber, was something that was not under human control. Using Krieger as support, Pickering argued that physicists deal with the world as a field of agency with material dimensions and the scientific world is amply stocked with material agents.[5] Human agency and material agency interact as "a dialectic of resistances and accommodations" in which machines are intermediaries that capture material agency as a particular combination of particular elements that acts in a particular way. The machine "is the balance point, liminal between the human and the nonhuman worlds."[6] Science is a collection of powers, capacities, and performances that achieves expression in "captures of material agency" through "whatever [is] required to set machines in motion and to channel and exploit their power."[7] It is through this process that material agency is determined in terms of temporally emergent "captures", whilst human agency is temporally emergent as "discipline". Machine and human performances occur simultaneously and, consequently, in experimentation, the process of this capturing material agency is one of "tuning" both human interventions and machine performativity in feedback relations with the other. The "constitutive intertwining" between human agency and material agency is continually and dynamically undergoing production through this dialectical process. It is practice that intertwines the contours of material agency with modes of human agency in such a way as to inextricably mix them together as an ontologically and epistemologically productive "impure dynamic". Human agency and intentionality are transformed and restructured throughout the process of trying to stabilise precise material configurations and "captures of material agency". Physicists may, in the process of trying to succeed in achieving any original goal or project, end up succeeding in a different goal or project because plans and goal are revised and are subject to mangling in practice. Accommodations take the form of adjustments to intentions and practices, the adoption of alternative techniques, changes in the material configuration of the apparatus, or employing expertise as a resource. These accommodations transform the original intentions. Human intentions operate in a field of existing machines in such a way that the goals of scientific practice are emergent in relation to this field as they take advantage of prior captures of material agency. It is this relationship between disciplined human intentionality and machines, in which both are mutually modified through reciprocal tuning, which keeps human intentions intertwined with material agency. The contours of material agency emerge as resistances to human agency because, without human agency, these contours would not exist. Material agency is emergent in the form of resistances to human intentions which, in turn, are modified, transformed, as accommodations to material agency. This was a central to Pickering's whole argument because

> [t]he resistances that are central to the Mangle in tracing out the configurations of machines and their powers are always situated within a space of human purposes, goals, plans; the resistances that Glaser encountered in his practice only counted as such because he had some particular end in view. Resistances, in this sense, exist on

the boundaries, at the point of intersection, of the realms of human and nonhuman agency. They are irrevocably impure, human/material hybrids, and this quality immediately entangles the emergence of material agency with human agency without, in any sense, reducing the former to the latter.[8]

Pickering proposed that scientific culture is a patchy, scrappy, disunity of diverse cultural elements in which scientific practice is nothing more than making associations between these elements. Material agency only arises as a result of scientific exploration finding new problems that arise when new machines are constructed to solve these particular problems. Pickering argued that the conceptual and material elements emergently arranged together and associated with representations through practice, in the form of "conceptual chains" linking representations and captures of material agency, constitute the totality of articulated scientific knowledge.[9] It is through this interweaving of conceptual and material elements that a representation of the experiment is constructed and linked within a pre-existent field of scientific knowledge. Concepts allow machine performances to be linked together with representations, spanning multiple levels of theoretical abstraction, and are emergent through practice. Pickering argued that the conceptual chains and representations associated with different machine performances are incommensurable with one another. For example, he claimed that post-1970s and pre-1960s particle physics have a different collection of machines and instruments and, therefore, the conceptual chains and representations produced by these phases of physics emerge from different practices and are incommensurable with one another.[10] For Pickering, this implies that the captures and contours of material agency are also incommensurable, given that the associations between cultural elements emergent from the dialectic of accommodations and resistances are different, and therefore have distinct "temporally emergent" ontologies. As a consequence of this, Pickering claimed that the content of scientific knowledge was nothing more than a series of "temporally emergent" concepts and representations associated with interactive stabilizations situated in a multiple and heterogeneous space of machines, instruments, conceptual structures, disciplined practices, and social actors. Technical knowledge, abstract laws, expertise, models, experiences, techniques, machines, concepts, etc., are elements, resources for mangling, that cannot transcend the dialectical process of experimentation and guide it. His interpretation left nothing that can help us to make associations because there is nothing outside "the Mangle" and literally nothing that could enable us to build machines.[11] It is all just happenstance. We feed our intentions in, "the Mangle" mixes up a load of heterogeneous cultural elements, transforms them, and spits out a product as an element for future mangling. If that product fulfils our intentions then that is simply a matter of good fortune. He explicitly rejected the notion that there could be transferable skills, any general knowledge of machine building, experiences, or even guidelines, which could enable us to build machines. All of these are nothing more than heterogeneous elements for mangling; they do not shape the mangling process. Pickering asserted that this "open-ended extension" was something that, in principle, could not be explained because

> Nothing substantive in scientific culture or anywhere else... necessarily endures through and explains the process of cultural extension; everything in scientific culture is at stake in practice; there is nothing concrete to hang onto there.[12]

In Pickering's analysis, the notion of "temporal emergence" meant that there is no substantive explanation to be given for the extension of scientific culture and, in the final analysis, "things just happen".

I agree with Pickering's claim that intentions, conceptualizations, and representations of machine performativity, and the available heterogeneous elements in experimental physics are produced, transformed, and interconnected by actually building machines and performing experiments. This is a fairly uncontentious claim. I also agree that the concepts and representations that link otherwise unrelated machine performances are produced through the real-time practices of experimentation. However, Pickering neglected to examine whether scientific products are fed-back into other projects in such a way as to actually *inform* and *constrain* the scientists involved about how to proceed with the project and how to connect otherwise distinct material practices within research. He did not allow the technological framework of research to actually guide and shape the process of mangling in any way, except as mere happenstance and opportunistic uses of prior "captures of material agency". Furthermore, by defining the ontological disclosures, emergent through experimentation, in terms of the interactive stabilization between human and material agency, Pickering's interpretation is somewhat positivistic and has failed to examine the metaphysical foundations that heuristically associate scientific models, via a concept of "natural mechanism", with those conjunctions of events. Hence, Pickering is unable to explain why and how experimental physics is culturally considered to be a natural science. Pickering was correct to identify the ontology of particle physics as the emergent product of modelling machine performances, but he misunderstood how models are used within human interventions to connect otherwise unrelated machine performances. His claim that the pre-1960s and post-1970s phases of particle physics have different ontologies and are incommensurable is only possible due to his positivistic conception of epistemology. Such a conception makes particle physics unintelligible. Yet, as I argued in the last chapter, once we examine how technological objects achieve transfactuality then we can readily see that this is not the case. In my terms, the mechanical realist metaphysics epistemologically unifies both phases of particle physics and allows both to participate in the disclosure of the same ontology. Thus overlapping strata of distinct machine-kinds are methodologically unified by the same metaphysics that defines the modes of disclosure of the ontology of physics in terms of the ongoing innovation of machine-families. Both phases of particle physics share some of the same components (electromagnets and ionising fluids, for example) and the models (electromagnetic theory, solid state physics, etc.), utilised to interpret machine performativity, shared some of the same theoretical indices (i.e. charge, spin, and mass). In order for these two phases of particle physics to be incommensurable they would have absolutely distinct meanings for these terms to have nothing in common whatsoever. Technological objects are transfactual and commensurable when they are used in different projects within the same technological framework to link phenomenologically different machine performances under the same theoretical description by using the same exoreferences, such as "electron", to bring together sets of exoreferenced functives, such as "charge", "mass", and "spin", to unify otherwise distinct machine performances via an exoframing model. It is through the same technological framework of overlapping machine components, models, mathematical techniques, etc., that distinct experiments can be said to all participate in the same unitary methodology. If projects share technological objects then physicists can transfer their exoreferences and exoframes between projects and build up a stock of experiences, resources, and tactics, a standing-reserve, which can be analogously used as possible accommodations in analogous projects. Although this does not guarantee success, it does mean that physicists can develop a bounded technical rationality, which can constrain and guide their intentionality in making selections of possible accommodations and identifying potential resistances. As I argued in the last chapter, models are implicitly involved in the construction of the apparatus, the development of operational procedures, and the making of observations, and modelling is crucial for the labour processes of experimentation to be an intelligible means of the disclosure and implementation of natural mechanisms. Models provide the transfactual interpretations between otherwise unrelated machine performances, providing commensurability, and are used to guide technological innovation by analogically connecting machine-kinds within the development of a metaphysically conceived technological framework. This explains how physicists represent their ongoing technological activities as a stratified process of achieving ontological depth, and, through the analogical and metaphorical use of models, also shows how technological objects are transferable between experiments, achieving autonomy, transfactuality, and commensurability.

Bounded technically rational choices can be made against the paradigmatic background of clusters and constellations of standing-reserve, expectations, and teleological positings. By allowing scientific practices to share stable cultural elements it is possible that physicists can acquire a cultural stock of technological objects, which enables them to choose particular accommodations as possible solutions to particular problems, and situates them within a shared technological framework. Bounded technically rational intentionality constrains and guides choices by associating and interconnecting machine-kinds with models and productivity in the design, construction, operation, and interpretation of new machines. Pickering did not allow this possibility in his analysis and, consequently, the choice of particular accommodations that particular experimenters made to deal with particular resistances is inexplicable to him except as *ad hoc* tinkering. Take Morpurgo's experiment for example. The technological background, with its associated expectations and techniques, not only permitted Morpurgo to postulate this experimental apparatus as a means to search for free-quarks, but it also constrained how he could use the machine to proceed with this search. It is this notion of *constraint* that is absent from Pickering's analysis and, consequently, Pickering could not explain Morpurgo's choices of the accommodations to emergent resistances during the process of building the experimental apparatus. For example, Pickering claimed that

> ... Morpurgo found that the charges on iron cylinders seemed to drift overtime – from zero to e/10, for example. *Tinkering once more in material practice, Morpurgo found a new way to frame material*

agency, discovering that he could achieve stable measurements, again of zero charge, if he spun the cylinders.[13]

By treating this new technical practice of "spinning the cylinders" as "a new way to frame material agency", Pickering ignored the extent that "measurements of charge" are techniques which are situated in a pre-existent technological framework that transcends and orders the particularities of experiences and practices. What led Morpurgo to consider "that the charges of iron cylinders seemed to drift over time" to be a problem? And, what led Morpurgo to consider spinning the cylinders as a possible solution? Pickering answered the first question by appealing to "resistances of material agency", and could not answer the second question except by appealing to "ad hoc tinkering". Although he did account for why Morpurgo had to make an accommodation, he cannot account for why Morpurgo chose the accommodation he did except by explaining it away as "tinkering". Why didn't Morpurgo try slaughtering a chicken and dripping its blood over the apparatus? It might have worked! I would suggest that the reason why "spinning cylinders" appeared to be a possible solution, and "ritual sacrifice" did not, was because the bounded technically rational choices available to Morpurgo were constrained by the technological framework in which his experiment was situated. The drift, as a resistance, was a product of expectations and judgements of what good measurement techniques and experiences would have been in that context. Morpurgo only "tinkered" in this way because he was using an interpretive model of why there would be charge drift on the iron cylinders and what he could possibly do about it. This model was inherited along with techniques, component machine-kinds, exoreferences, exoframes, and other technological objects, through the extension of the analogical model within the technological framework. The spinning, as a possible solution, an accommodation, was more a product of the Morpurgo's expectations, technical experiences, and theoretical interpretations of "charge distributions" and "the properties of iron" than it was "mere tinkering".

According to Pickering, material agency emerges when human beings *actively* construct a new machine and *passively* observe the performance of the machine to see whatever captures of material agency occur.[14] This is apparent in Pickering's analysis of Morpurgo's observations when he wrote:

...Morpurgo assembled his apparatus, switched it on, and then, *surrendering his active role*, stood back to watch what would happen –

literally, through a microscope. Swapping roles, *the material world was in turn free to perform as it would*; the grains levitated and moved away from their equilibrium position when the electric field was applied.[15]

However, by claiming that there are two distinct and identifiable phases of scientific work, actively constructing the apparatus and then passively observing the results, Pickering has built the human-material distinction into his analysis and neglected the way that experimenters, like Morpurgo, *simultaneously* and *ambiguously* "actively" and "passively" perform experiments. The "active" choices that Morpurgo made in the construction of his apparatus were simultaneously "passive" responses to what was the bounded technically rational choice, according to his current expectations of the alethic possibilities within the technological framework. The "passive" observations that Morpurgo made, after switching the apparatus on, were simultaneously "active" interpretations and construals of what he was seeing, made in terms of his current model of how the apparatus worked (i.e. that an electric field was being applied and how this should effect charged grains). Pickering was aware that Morpurgo used an interpretive account of how the apparatus worked in order to "move from observations of the response of the grains to an applied electric field to statements about the electric charges carried by the grains."[16] However, Pickering neglected to attend to the extent that an interpretive account is also required to make observations of "the response of the grains to an applied electric field" in the first instance. Otherwise, why would we say that there was "an electric field" present and it had been "applied" by turning a dial, pressing a button, or whatever procedure we associate with the technique of "applying an electric field"? Making observations involves simultaneously passively/actively interpreting what is happening during the experiment as interventions, observations, and expectations ambiguously interact. This is apparent when Morpurgo observed an anomaly. He was frequently confronted with anomalous grain motion in terms of his current expectations whilst attempting to make sense of it in bounded technically rational terms. When he could not, he had to adopt an alternative tactic, which Pickering characterises as "active", by modifying the formula for calculating charges on the grains by adding an additional term to the equation. However, contrary to Pickering, this tactic was also "passive" because the choice of techniques for the active modification of the formula part of the model is constrained in accordance with what he perceived the bounded technically rational choice to be in relation to what Morpurgo could measure using his apparatus, how he could expect to exoreference that modification as a physical mechanism, and simultaneously what he could expect to demonstrate by exoframing the apparatus in that way. The configuration and functionality of the apparatus, as well as his expectations of the functional and demonstrative potentials of his models, interpretations, and the apparatus was constructed and exoframed within the constraints of current bounded technical rationality.

Goals and expectations are an intrinsic part of any technology and are a pre-requisite for bounded technical rationality. Without goals and expectations we could not make judgements about which techniques, tools, or machines to use, and we would not expect them to work or break down in particular contexts and perform specific productive acts. Bounded technically rational choices, decisions, and intentions are constrained by technology to the extent that the existence of particular technologies are ontological pre-requisites for the existence of particular choices, decisions, and intentions. The teleological positings of labour bridge practices and experiences, revealing the extent that both are intrinsically defined in terms of each other, in accordance with models of efficiency and functionality, during productive processes within the current technological framework. Goals and expectations are emergent in the same way as any other technological object and cannot be localised in purely the human realm. Glaser could not have intended to build a bubble chamber if he were born into Galileo's culture. Nor could Morpurgo have intended to search for free-quarks by sacrificing a white ox at Stonehenge during the summer solstice. Neither would have been a bounded technically rational choice within the historically emergent technological framework. This constrains intentionality through bounded technical rationality because a physicist has a technological constraint placed upon him/her to use only theories, interpretations, and intentions, which contain measurable elements and the choice of theory, interpretation, and intention is constrained by the limitations of the available measuring technologies. In the context of experimental physics, the physicist is constrained in her/his choice of technological objects according to the consensus of other physicists regarding what can or cannot be demonstrated by using particular techniques and machines. The physicist is constrained by the template of methodology and the paradigmatic consensus regarding "efficient techniques" and "achievable goals", and the bounded technically rational choice is always to choose these "efficient techniques" as the means to achieve "achievable goals". However, the technological imperative to innovate is central to experimental physics because it challenges human agents to innovate solutions to technologically produced problems and to explore the ontology of the world by increasing the available technological possibilities. Not only does technological innovation constrain, extend, and challenge intentionality by creating the possibilities for a whole range of potentially achievable new projects and goals, but also constraints and challenges intentionality when human agents maintain a perpetual deference to the future, regarding the perpetual revisable estimations of appropriateness and efficiency of any technique, as well as the achievability of any goal. The "bounds" of technical rationality are themselves at stake and tested by experimentation and innovation. Thus Glaser and Morpurgo were challenged to invent their machines in accordance with the goals, expectations, and constraints of the technical framework in which they were situated and embodied, as a test of the alethic possibilities and limitations of that framework. They were destined as soon as they took up these challenges.

Pickering required a concept of "material agency" because he left no space in his analysis for constraint and challenging within a transcendent technological framework that pre-exists and orders the particularities of experiences and practices of experimentation. Thus he maintained that the experimental apparatus, the machine, is an intermediary for the accommodations and resistances of the dialectic between human and material agencies. However, if we restrict our analysis of experimental physics to its performances, as Pickering implored us to do, then we do not have a phenomenological experience as machines as intermediaries except in terms of what is said about them. What we have are machines and interpretations of their performativity in terms of mechanisms, laws, and theoretical entities. It seems to me that we describe these machines as intermediaries between human agency and material agency only because such a way of discourse is a product of removing the machines from the account and replacing them with Nature (or material agency in Pickering's case). If we examine machines *qua* agents rather than intermediaries then we can develop a rather different description of experimental physics than Pickering does. "Material agency" is an abstraction of the particularity of practices and experiences emergent through complex labour processes that are challenged to produce the knowledge of general principles of mechanisms and functions of mathematically abstracted machines, in order to metaphorically make the natural world intelligible to human beings in terms of mechanisms and functions. The functionality of any machine is defined in terms of the larger organizational structures in which they are integrated in accordance with a teleologically posited goal. The experimental connection of teleologically posited machines with other machines generates "emergent brute resistances" in accordance with the degree of productive incoherence within the technological framework. This "resistance" occurs when productive agents interfere with one another and is the consequence of their heterogeneity. Coherence is achieved and resistance disappears when these divergent heterogeneous agencies are brought together upon the anvil of practice, so to speak, and forged into a single centre of homogenised agency. It is the labour process itself that transforms heterogeneity into homogeneity by teleologically transforming the particularities of fragmentary experiences and practices during the processes of integrating them within a general and unified technological framework.

Human agents and machine agents are distinct but inseparable from one another in the technological contexts of experimentation and innovation. Machines are products and embodiments of intentions, expectations, beliefs, choices, and values, and as such are agents within a society. The innovation of new machines, when used, generate (not just transform) new intentions, expectations, beliefs, choices, and values, through the powers, constraints, challenges, and demands, that they produce and disseminate through the society in which the are operational. Machines teleologically shape the productive directions and possibilities of labour and, it is in this context, that machines can be said to have agency within their own right. For example, in Glaser's work, by attempting to connect diverse components together with the aim of constructing a unified machine, a particle detector, each of those components began as its own centre of agency (as a result of previous unification within the work of others) and the problem Glaser faced was bringing these diverse agents together into a coherent and unified totality. The resistances arose through the problem of achieving stable co-operation of functionality as each component was connected together. When these heterogeneous components were brought together, the particularities of each functioned out of phase with the others, and the outcome was incoherent with the teleological posited goal of producing a particle detector. This divergent functionality is a consequence of components being brought together to perform functions, for which they were not previously designed, developed, and stabilised to perform, within a novel and incomplete framework. This incoherence cannot be identified with an "emergent material agency" precisely because it arises due to the fragmentation and disharmony of divergent particularities rather than from a unitary source. It is a product of non-linear heterogeneous agencies that find their functionality through their application upon each other in accordance with the teleological positings that provide principles by which they are ordered. Hence, Glaser had the problem of having to coherently integrate his bubble chamber into the wider context of particle physics. This involved connecting his machine, itself a componential complex, as a component in the larger technology of particle detection. Achieving stability involved not only integrating all the componential agencies of the parts of his machine but also integrating his machine, as a component, into particle physics. His work required the integration of diverse organizations of agencies that involved not only integrating electrical components, glass tubes, and strange gases, but also involves integrating techniques, interpretations, conceptualizations, political institutions, economic factors, social organizations, beliefs, values, and expectations, together into a stable means of production. It is only by doing this, can new machines, like bubble chambers, be made to work as agents and become part of scientific culture.

This is especially apparent when we examine how materials are actually used in scientific work. Materials are integrated into the construction of machines in order to determine their properties on the basis of what that material does, as a component, to the other components, the other materials, to which it is connected and acts upon. From the onset, this intervention is bound-together with representational models of the material and the machine. Each component is organised within the technological framework, teleologically posited in terms of functionality, to explain the structure of the machine in which it is a component. After all, what is a machine? It is not merely a particular configuration of materials (metals, plastics, glasses, gases, etc.) but it is a particular configuration of functions that can only be emergent within contexts of implementation. Machines are made to reproduce functionality through integrating diverse agencies into a coherent unity, and, as such, may be mechanical, mathematical, computational, social, political, military, biological, medical, scientific, analytical, sexual, etc. Functionality does not necessarily come out of some inner principle of Nature, but it does require organised power to set-upon otherwise heterogeneous objects, gather them together, order them, work upon them, and integrate them into a stable, coherent, and unitary function, in accordance with the teleological positings emergent from an embodied heterogeneous complex of diverse modes of agency. It takes effort, resources, and power to create, unify, and maintain new orders of functionality and purpose within a perpetually transforming technological framework. Complex teleological labour processes driven by the ongoing challenges of *Ge-stell* make functionality. Even the so-called basic "materials" from which any machine is built are functional components with their own histories of stabilization and implementation. This is even true of substances that we take for granted as elements, but understand in terms of a long and protracted history of use. Take iron for example, this substance is itself identified in terms of its functions of hardness, durability, tensile strength, availability, cheapness, etc., and all these functions have taken considerable work to organise into a coherent unity. Iron ore is dug up out of the ground, but iron is made and understood in terms of its appropriateness and resistances within the context of a long history of heterogeneous teleological positings and labour struggles to achieve them. Resistances can be accounted for by examining the structures of intentionality and appropriateness in terms of means-ends relations, in the dynamic interaction between heterogeneous modes of agency, which may cohere or incohere with each other, and, consequently, dynamically produce stable or unstable intentional structures within wider and pluralistic contexts. As such, resistances are the result of sociological, psychological, political, economic, and technical incoherence between divergent and competing agencies, each a componential complex with its own history of development and future of transformative change. By treating "the source" of such resistances as the interaction of human agency and material agency, Pickering has substituted "material agency" for the wider context of the outcome of complex teleological efforts and struggles to integrate heterogeneous innovations into pre-existing technological and social orders. These efforts are inherently unstable because their non-linear interaction mutually transforms their functionality. Contra Pickering, the situation is more one of flux than dialectic and the challenge of invention is to create, unify, and maintain order and unity of purpose and function from out of the fragmentary and pluralistic chaos of non-linear technological heterogeneity.

Whence the resistance? The problem facing us with this question is how do we locate and identify the source of particular resistances, in its unity, within contexts of heterogeneity. How can we make sense of the possibility and phenomenology of resistance without adopting scientific realism? In the Aristotelian scheme this hindrance and resistance is an example of *hyle* rather than *phusis*. The term *hyle* captures both the sense of resistance, or recalcitrance, and also the sense of the way that appropriateness emerges dialectically from ongoing interventions and intentions. It is the phenomenological particularity of the particular and cannot be generalised between different contexts of productive activity. It is a phenomenological response to human interventions that does not spontaneously come of itself. *Hyle* is emergent as a consequence of the attempt to impose the idealised form and intentions upon materials and the particularity of that deviation from expectations is dependent on the content of those expectations. It is an emergent feature of practice that occurs during the human attempts, guided by *techne*, to inscribe form into materials, which is neither controlled by human intervention, nor does it exist independently of human intervention, but it is a property of the context of *poiesis* guided by *techne*. *Techne* is an imagined complete and invariant knowledge of what is considered as "a true course of reasoning" involved in bringing something into being through a communicable and universally repeatable act of *poiesis*. In physics, the universalisability of any such act would depend upon its successful integration into a machine prototype and its successful utilization in the extension of a machine-family or the innovation of a new machine-kind. Invariance, as the repeatability of a result through the repetition of intentions, interventions, material practices, and the "true course of reasoning", is itself a techneic and poietic reproduction of the transitive process from which it came about. A techneic theory is one that is used to imagine a specific object utilising the same cluster of technographe, techniques, material practices, and machines. *Techne* is imagined to provide the contours, boundaries, limits, and prescriptions, to a set of agential possibilities that are indexed in terms of causal mechanisms and their effects. As the knowledge of the Being of Becoming, *techne* aims at unifying any particular practices, by generalising the intransient causal principles of change within that work, and offers complete alethic knowledge as "the end of experimentation". However, no such knowledge is actually possessed outside the imagination and the complete unification of practices and experiences remains incomplete. Complete and perfect reproduction is an imagined ideal that is never achieved in practice and experience. *Hyle* demonstrates that extent that the technological framework is incomplete and the particularities of the particular deviate from the law-like behaviour.

Oersted's famous experiment with a magnetic needle, a chemical battery, and a wire, provides a relatively simple example of phenomenological emergence of *hyle* in response to human interventions. As Oersted reported in 1820:

> ...The opposite ends of the galvanic battery were joined by a metallic wire, which... we shall call the uniting conductor or the uniting wire... Let the straight part of this wire be placed horizontally above the magnetic needle, properly suspended, and parallel to it... Things being in this state, the needle will be moved, and the end of it next the negative side of the battery will go westward...[17]

Oersted proposed, on the basis of this discovery, that an electric current causes magnetic effects. However, it is not easy to reproduce this effect. The needle does not move in a clear and stable way. It is rather chaotic and it is difficult to witness the reported effect and keep the needle from touching the wire. Gooding termed this as "the recalcitrance of nature" and it resonates with Tartaglia's "material hindrances" and Giovanni di Guevara's "marvellous motions", in that they were all referent to the particularity of the "natural" response of artificial devices to particular human interventions. It is this phenomenologically spontaneous and chaotic behaviour that Pickering termed as "material agency". It is an abstraction of the phenomenological experience of machines "doing their own thing" – they way that their performances do not cohere within the technological framework – and is not a pole or terminus of a dialectic. It is only through particular conceptual structures, socio-technically constructed and inherited, on the basis of metaphysical presuppositions, that notions such as "material agency" can be intelligible as a conceptual device to explain the phenomenological experience of resistance when bringing together divergent modes of agency within a teleologically posited labour process. It requires further rhetorical and analytical work to abstract the machine performances into an intermediary and the complex labour processes of experimentation into an imagined dialectic between two modes of agency. Pickering oversimplified building machines by analysing it in terms of human agency and material agency. In my view, we cannot make the process of building machines intelligible by analysing it in terms of dialectic between these two types of agencies. Building machines involves integrating a diverse and heterogeneous set of complex modes of agency, within the context in which any machine is a particular framework of interactions that interacts with other modes of agency, within the environment in which it is contained. Machines are technological objects within larger technological frameworks that order each other and, in turn, are ordered within processes of gathering and ordering standing-reserve to satisfy the never-ending challenges of *Ge-stell*. Rather than demonstrating or indicating a lack of correspondence with natural laws, resistance is emergent as an incoherence within the unfinished processes of labour and is contextually conceived as a "malfunction" or "problem" until the coherent functionality of all agencies are produced and unified as an outcome of the innovative processes of labour in the creation of its own possibilities. What is the source of resistance? *Why do we think that there is a single source?* Resistance arises from heterogeneity, diversity, and incompleteness.

What enables us to build machines?

What enables us to build machines? In many ways this is the central but unasked question of experimental physics as a whole. Or, to put it another way, the construction of experimental physics is itself a performative attempt to answer this question by building machines and attempting to present causal accounts of how they work. The presupposition of the existence of natural laws in response to the question does not answer this questioning. For even if there were natural laws we still would not know why there are natural laws at all, or why they have the form that they do. Experimental physics, as a mode of *Ge-stell* challenged to achieve its own *techne*, is an ongoing process of producing causal accounts of the technological innovation of prototypes as technological objects for further innovation. The challenge for experimental physics is to explore every conceivable possibility of its own destining. It cannot end until it has undertaken every challenge that it sets upon itself. In other words, the task of experimental physics is to design, build, operate, innovate, and perform every possible experiment upon every possible machine-kind. However, even if we imagine this to be a finite task for which completion is a possibility, this still raises the question of whether experimental physics will ultimately provide the answer it was set up to provide. Whilst the objective of experimental physics remains the disclosure of mechanisms, through innovating novel machines, it will not answer this question. Why? Physics is challenged to build machines to explore the laws and mechanisms in operation upon the interface of machine performativity. Thus it is destined towards the question of what enables machines to be built *as if* the answer was itself something mechanical. However, what physics cannot address is the being that builds machines in order to understand how they work. It cannot address the "us" of the question. This is more than just a question of how the emergence of the being of *poiesis* through the processes of machine agency in a technological society is possible. The question of what enables us to build machines is a question of the origins of the possibilities and conditions of the processes of labour themselves. It is a transcendental question of the metaphysical processes of artifice, labour, and productivity. This is a question of the being and becoming of "us". The mystery at the heart of the teleological labour process is the mystery at the heart of our being-in-the-world. Why is the world like this? Why are we able to act in the world in the ways that we do? Why are we like this? Who are we? What are we doing? Why are we doing it? These questions cannot be answered by modern physics because it cannot address the mode of being for which its causal models are intelligible explanations. It cannot address the question of its own metaphysical presuppositions. The models produced by experimental physics are techneic and only relate to the parts of the world that are contained within the processes of experimental physics, that at most, can only reveal how a part of the world works. That part of the world is restricted to the machines involved. Even if we accept that physics can successfully (eventually) produce knowledge about the first and necessary causes at work, in such machines, it does not follow from this that this knowledge is applicable to any other part of the world. It certainly does not follow that this complete set of mechanisms will somehow explain the entire "worlding of the world". Nor does it follow from the fact that experimentalists claim to discover and use natural mechanisms, with considerable social and technical success, that any such mechanisms exist in a realist sense. A mechanism, as an index for a reproducible change, names a whole complex of reproduced ensembles of machine performances, techniques, and exoreferences. It is a conceptually unifying index for a complex of technological objects that can only be identified through its implementation as a transferable abstraction within the social imagination and technological practices of a technological society.

From an outsider's perspective we can understand the motives of physicists without uncritically accepting their metaphysical presuppositions. Providing we attend to how physicists use these presuppositions in the construction of their practices and experiences, then we can explain how they produce explanatory power and predictive success. Physics does not need to correspond to any reality outside of itself. It only refers to Nature rhetorically and poetically as a metaphor.

When a physicist proposes natural mechanisms to answer questions of how natural phenomena come to be, why stars shine or birds fly, for example, s/he is proposing a template for a techneic method to explain how s/he could implement these mechanisms to make an artificial star shine or make an artificial bird fly. S/he makes a model of a machine as her/his answer. Even if physicists could make shining stars or flying birds, which presently they can not, they still would need to assume mechanical realism in order to claim that there was only one way to do it, and they knew how Nature did it. Of course, as has been argued throughout this book, experimental physics does presuppose the mechanical realist ontology, but it does not follow from this that the ontology presupposed by experimental physicists is definitive of the ontology that they actually explore, or that the world is exhausted by the ontology of the experimental sciences. At most, experimental physics can only epistemologically justify a modest mechanical realism by revealing their capacity to model productivity in terms of the enduring mechanisms at work in machines.

Whilst Bhaskar did provide us with an internally intelligible account of experimental sciences as explanatory natural sciences, he did not provide us with any reason why we should believe that mechanisms occur in open systems at all. By the qualification that they are repeatable, a quality denied by Bhaskar to the phenomena of open systems, there are good reasons why mechanisms, even by Bhaskar's account, could be taken to only occur in closed systems. However, my argument has been that mechanisms are ontologically restricted to the kinds of machine-families and productive contexts in which they occur and epistemologically extended as metaphors to provide causal explanations of natural processes. They are technological objects that are given epistemological significance on the basis of metaphysical precepts and do not necessarily occur in Nature at all. They are the products of a metaphysical performance art that is itself empowered through its non-linear power relations within the wider world of social, economic, and military goals. Bhaskar's claims go too far. His argument only shows that experimental physicists, aiming to discover causal laws, cannot be empiricists or idealists. Nothing else. Bhaskar's transcendental realist ontology depends on a reified technocentricity and a concealed metaphysics that could only be described as referent to Nature if "Nature" is taken to be that which is revealed by the technologies of experimental physics. His claim to maintain a "nature-centricity" is a form of *anthropocircumferentialism* based upon a conception of "technical man" as a natural being.[18] However, once the technological character of physics has been addressed then the realist notion of "causal power" is open to the criticisms that it is a reification of those complex labour processes into abstractions. It is the reification of the technological society as neutral, normal, and natural. The interrelationships between technological objects, within the wider world in which they are situated, need to be analysed in terms of *complexity* rather than causality. On my account, the primary relationships between scientific discourse and "Nature" are reproduced relationships of power and agency situated within social imagination as a mechanistic world-picture. The scientific attempts to identify "truth", "efficiency", "natural law", and "causality", are attempts to reproduce techniques, interpretations, practices, values, dogma, institutions, orthodoxy, and authority, through acts of closure and the exercise of social power, in the face of contingency, plurality, controversy, and chaos. They are attempts to impose order upon the world. Science, technology, human agency, and human experience (in the technoscientific cultures and societies of the current era) permeate and penetrate each other to such an extent that it is impossible to separate them. The relationships between technoscience and the world in which it is situated are understandable in terms of non-linear "feed-back" loops in which both are defined and transformed in relation to the other. Technoscience legitimates and circumscribes technological conceptions of "Nature" whilst being legitimated and circumscribed by the technological society, which, in turn, uses these conceptions to legitimate and circumscribe metaphysical conceptions of the human condition and our future possibilities.

Technologies are transformed when they coherently interact with one another into a new unified mode of productive agency. The emergent functionality of any machine can only be contextually determined in relation to the ensemble in which it has been integrated as a component, and the innovation and integration of any novel machine in an ensemble of other machines transforms the productive context of that labour process. Machines realise and exercise their transformative powers during interactions with other technological objects, transforming them and being transformed, and, therefore, each and every machine performance is the product of a non-linear ensemble of interacting technological objects in which each is emergent from a background complex of heterogeneous transformative powers. The greater the numbers of technological objects brought together in different contexts then the more combinations of transformations are possible. Transformative powers emerge as the consequences of implementing any teleological posited goal within a series of experimental interventions that comprise the innovative labour processes in accordance with heterogeneous demands and challenges of a wider society. Ontological depth is the use of subsequently developed strata of machine-kinds within a process of "reverse engineering" to transdict the performativity of earlier strata in terms of the later. This is a mode of ontological extension through innovation that is metaphorically used as a mode of explanation of the technological innovation of novel strata. It is my proposal, posited as an alternative to mechanical realism, that the innovation of novel strata can be understood as a creative process. Realism presumes that the world is complete and every possibility is determined in advance. We have no way of knowing this. I have argued that experimental physics does not require such knowledge, or even its possibility, in order to progress. Each innovation of a new stratum is a revolutionary moment. It is a creative event. These are singular moments in which it is impossible to determine whether the novel prototypes will disintegrate into chaos or integrate into a new order. The creation of any new order emerges from the convergence of otherwise distinct clusters of technological objects and a unification of their associated powers. This is a process of mutating trial and error that spontaneously generates a novel shift in the ordering process in which heterogeneous objects are combined together under the same index. In any complex process, new levels of complexity can be achieved (almost randomly), which cannot be understood in terms of the previous levels because their indices, rules, and abstractions do not apply. A stratum of new rules and abstractions are in operation as a new organization is brought into being from the old and is irreducible to the former. There is no reason to presuppose that the possibility of successfully achieving this ongoing stratified productivity has a unitary source. There is no single unitary source of transformative powers because they can only emerge because the pluralistic, unpredictable, and non-linear heterogeneity of technology. A single source could only generate homogeneity and linearity; consequently, the innovation of novel irreducible strata of novel techniques and prototypes would be impossible. In a complex process it is the incompleteness and disunity of the process that creates the non-linearity of innovation, because what is brought-forth partially depends on how it is brought-forth, what it is brought forth into, and its posited purposes. It is the whole technological society that is experimental.

The distinct kinds of mechanisms proposed to explain different kinds of novel machine performativity will create novel levels of complexity for which new rules and abstractions are required. The claim for the pre-existence of laws to explain complexity is premised upon a deterministic and linear conception of evolution. As an alternative, we can attempt to analyse this process as a non-linear creative process of ordering heterogeneous technological objects and attempting to explain their unification in terms of an ongoing process of innovation. For example, the "neutrino oscillation" has been proposed within contemporary astrophysics and particle physics to explain the deviation between theoretical expectations and observations, but its successful implementation of it within experimental physics depends on more than its utility as a transdiction. The "acid test" of any transdiction is not only whether it provides an intelligible causal account of the anomaly in question, but whether it provides exoreferences and functives for the exoframing of further technological innovation. One of the prime difficulties for quantum theory, for instance, is that it provides functives and facilitates innovation but does not provide causal accounts. However, the instrumental success of quantum theory is sufficient for it to be continued as a part of mainstream physics until an alternative theory, which also provides explanatory power for the same kinds of machines can be found. However, it does not follow from the successful satisfaction of this "acid test" that any such theory is (a) true of Nature or (b) the ultimate explanation. Neither technological utility nor explanative intelligibility is a necessary or sufficient criterion for objective truth about Nature. It is logically possible that objective truths about Nature are both technologically useless and incomprehensible. However, useless and incomprehensible truths will never be a part of physics, and any truth that is part of physics has been disclosed by building a machine. Once these processes of production and demonstration have been successfully invented and disseminated, from the standpoint of a scientific community, then t he interests of that community can move on. Necessity arises in hindsight through the successful production of acts of reproduction. The maxim that "necessity is the mother of invention" can always be countered with the maxim that "invention is the mother of necessity". The assumption of mechanical realism allows the term "artificial" to be dropped and the adjective "natural" to be inscribed in its place. Nothing more.

It is quite arbitrary to isolate a single component of the interconnected complex technological framework that constitutes the background for innovative socio-technical agency as being the single element responsible for its success. Predictions derived from a theory are only components of the whole process of exoframing and performing experiments. If we aim to understand any particular experiment in terms of its components then we must holistically examine the complicated and intricate interactions, from their set-up to their completion, in productive processes in which the teleology of the complex has ontological primacy over its components. The components involved in the whole complex can only be understood in terms of their total concrete interaction within the particular experiment in question as a single complex technological object. It always remains an open question of how these components were teleologically used and connected within the whole complex of the closed system. The task of attempting to mentally reconstruct the experiment in terms of isolated components is an endless task. Concepts, functives, material practices, social practices, visualizations, metaphors, machine components, measurements, calibrations, and models are all inextricably bound together in the design, construction, operation, and interpretation of machine performativity. These components can only be understood within the context of a non-linear analysis of production of particular machine performances in terms of the functions that they fulfil within the closed system. This involves understanding the whole process of constructing and interpreting the closed system from beginning to end. However, given that the meaning and functionality of any experiment is determined in relation to its interpretations and uses then we should not limit our analysis to the closed system. The task of tracing the function of any single component within any experiment requires a complete analysis of the whole experiment, within the open system of the wider world, to be set-up and performed in order to fully understand its meaning and functionality. All technological objects are complex interactions between other technological objects, within a technological framework that is situated in an environment, and, hence, we cannot treat mechanisms, capacities, or tendencies as ontologically fundamental without presupposing metaphysical precepts. Components must be understood in terms of their purpose and performance, as teleological and non-linear, within a labour process and the wider world. In this sense they cannot be understood as isolated components at all. The identification and representation of a single element as responsible for the success of any socio-technical agency is itself the invention of a tentative, hypothetical, and experimental object of experience and theory that is perpetually open to refinement and replacement.

Each component is an irreversible precondition for the agency of other components and, consequently, the being of the whole process is an irreversible extension of reality through the teleological and ontological challenges of *Ge-stell*. This is the reality of *poiesis* in the modern world and it is this reality that provides physics with the possibility of discovering the alethic possibilities of that world. However, the disclosure of *poiesis* is not the disclosure of what existed prior to the experiment but is the disclosure of what the experiment has brought about. It is the creative and innovative disclosure of experimental physics itself. This can only be determined *in hindsight* and the intellectual process of analysing an experiment is always one that attempts to understand what was actually done. Thus the understanding of physicists always lags behind the transformative extension of reality that they are challenged to bring about and the physicist cannot know what s/he is doing whilst s/he is doing it, because this can *in principle* be only known after it has been done because there is no "it" until it has been done. On this account, experimental physics is a mode of creative labour that brings beings into the world and transforms the world. The "it" is manifest through the contingent interaction of components and, to the extent that "it" is constructed as a technological object, only achieves "its" identity as a consequence of labour processes in innovative interaction with other labour processes. These technological objects are contingent upon both the paradigmatic background, against which they are contextually understood and used, and the teleological positings that emerge from the challenges of *Ge-stell*. As has already been argued, the causal powers or capacities involved in any complex process can be determined only from a position of hindsight using the reconstructive analytical imagination. Transformative power remains ontologically opaque and oblique to such epistemologically contingent and perpetually revisable causal explanations. These causal accounts remain only contingently and pragmatically associated to the transformative powers that the physicists bring-forth into the world. These causal accounts may well technologically function within the ongoing development of that experiment and the social use of those accounts to partially explain some aspects of the world, but they do not explain how transformative power is possible and brought-forth in the first instance. To put it into metaphorical language, the fire of the cave burns independently from the efforts of the shadow puppeteers and remains unexplained by the successful prediction of the sequence and changes in the progression of shadows. The way that physicists learn from and develop the process in their extension of their range of puppets and their refinement of the art of shadow puppetry tells us nothing about the flame itself. The shadows on the cave wall may well change as the puppeteers design, build, modify, and operate new puppets in the flickering light of the fire, whilst they carefully attend to the shadows and the flicker, but they cannot explain the flame itself. The shadows on the cave wall only relate to changes in the puppets and do not correspond to the fire at all. The prisoners in the cave and the shadow puppeteers cannot explain the fire.

Leaving the cave of the shadow puppeteers

Experimental physics is directed towards disclosing mechanisms and testing their reality by implementing them in the ongoing activity of its own technological innovation. The focus is on the expansion of possibilities and not upon histories. Due to the centrality of innovation within "testing" in physics, secured within a technological society, driven by the psychological desire for novelty, progress, and instrumentality, experimental physicists lose interest in their past very quickly. They have little interest *qua* physicists in what physics has been and are only concerned in the becoming of physics. They forget the tentative poeticising in the construction of metaphorical understandings of machine performativity. In this forgetfulness, they find their own truths. When it was forgotten that these models were "representative" then they could be presented as the real things. It is this habitualness, and forgetfulness, combined with an innovative headlong rush to discover new powers, that transforms metaphors into techniques, and techniques into truths. Physics is a heterogeneous convergence of machine families and metaphors perpetually directed towards its own *techne* as the asymptote of its expanding horizon. Physicists seek out heterogeneity to transform into prototypes of machines, models, and techniques. These strata are standing-reserve as resources for future innovation. Its end is the expansion of itself and an exploration of its own subtleties during that expansion. As an innovative process it is directed towards itself. To understand itself, in its own metaphysical terms, is to understand the Universe. In order to understand itself, to understand the Universe, it must explore all its possibilities and potentials. It must become everything that it can become. It must consume Nature and replace it with itself. Once it has done this then it will be its own truth. There will be nothing left to do, nowhere to go, and physics will simply cease to exist. All that will be left will be its technologies and no possibility of further innovation. It is a destining driven by the socio-technical organization of expansive innovation. It is creative of itself through poetical and technological innovation. It is an art that discovers the real by producing the real. It is a form of *poiesis* that transforms itself as its own resource and object. The ongoing process of *Ge-stell* is one that continually challenges physicists to order their practices into a concretely structured complex of inter-related mechanisms available for future innovation. However, due to the metaphysical precepts of experimental physics, the production of intelligible accounts of the causal processes at work in the production of natural phenomena according to natural law is the revealing of truths, in Heidegger's sense of *aletheia*, as a mode of disclosure for its own sake. Experimental physics is intimately bound up with *poiesis* as a craft and art and is also bound up with modern technology as *Ge-stell*. It bridges modern technology and craft practices and, as such, reveals Nature as both standing-reserve and truth. The production of technological objects is both a means and an end for further innovation. Physics is itself an experiment that began in the fifteenth and sixteenth centuries and it has been innovating itself ever since. By analysing experimental physics in terms of experimental labour processes we are able to address the fact that its successes and failures are the results of protracted historical struggles of heterogeneous organised efforts to stabilise and reproduce socio-technical practices involved in the design, construction, operation, inscription, and interpretation of machine performativity, whilst simultaneously situating that process within a perpetually refined world-picture. These efforts are always able to draw upon a background of prior successes and failures. In this respect the successes of experimental physics are not a "miracle" and are quite unsurprising. Or to put it another way, the successes of experimental physics are no more miraculous or surprising than any and every act of making.

As argued above, the truths and reality disclosed by the processes of experimental physics are brought forth as *aletheia* rather than mere "facts". Physics aims to disclose mechanisms and situate these within its ongoing activities rather than merely compare linguistic truth propositions with experience. Physics is situated upon the boundary between *techne* and *Ge-stell* because its alethic modalities are both goods in themselves and standing-reserve for future work. On this point I agree with Bhaskar's argument for an alethic conception of scientific truth (rather a correspondence notion) because the transitive dimension of human productive activity produces it. However, he did not fully escape the traditional epistemic correspondence notion of truth because he limited his conception of the significance of alethic possibilities by his commitment to the existence of an intransitive dimension of natural laws. The mechanical realist presupposes the truth of the premise that only the alethic modalities that correspond with the possibilities derived from "natural law" will work in the real world. Thus the creation of an artificial closed system can disclose "natural law" by exploring the productive possibilities within the closed system. The mechanical realist presupposes that only those artificial means that function according to "natural law" are capable of functioning at all. Human agents may well have only artificial means at our disposal but, for the mechanical realist, these means are only capable of being means in virtue of their utilization of "natural mechanisms". For the mechanical realist, the only technological objects that are possible are the ones that are constructed in accordance with "natural law". Hence, once mechanical realism has been presupposed, any possible artifact is not radically different in kind from any possible natural entity. The only difference between the two, on the mechanical realist account, is that the former requires human intervention to occur whereas the latter is the result of a lack of human intervention. They are merely counterfactuals from the same "natural laws". For the mechanical realist, such as Bhaskar, this is a requirement for the intelligibility of experimentation. This was presupposed in Bhaskar's subsequent philosophical writings, where labour and its possibilities are circumscribed and delimited in accordance to the possibilities permitted by natural laws; only that which is permitted by natural law can be brought forth because natural laws govern the conditions and possibilities of discovering, exercising, and realising mechanisms. Thus, for Bhaskar, human freedom does not consist in an independence from natural law but, rather, in the knowledge of natural laws and the possibility of making them dialectically work towards definite ends.[19] Supposedly the knowledge of natural laws and mechanisms provides the Promethean (and Marxist) promise of instrumentally increasing the material possibility of satisfying teleological positings that liberate us from the organically evolved alethic modes of our "natural state". Hence, it must also be recognised that, according to the mechanical realist metaphysics within the societal gamble, the technologist and the physicist are engaged in the technological imperative as a moral imperative. The gathering and ordering of all plans and activity in a way that corresponds to technology, within a technological society, is seen to take us nearer to the realization of the success of the societal gamble and the removal of the evils of the natural world. The drive to innovate, bringing novel inventions and new transformative powers, is inherently a moral drive within the technological society to improve it. This imperative imposes a duty upon trained specialists to fulfil their social responsibility to discover and utilise the most efficient means at our disposal and to create the means to liberate our fellow human beings of the arbitrary capriciousness of the vulnerability, disease, ignorance, and premature death that Nature imposes upon us in "our natural state" of animality. When Bhaskar made the ontological argument that human freedom is enhanced by the knowledge of natural law, his metaphysical argument for a realist theory of science was inherently a moral argument in which scientific truth is equated with the good. Thus, when Bhaskar equated increased freedom with increased techneic knowledge and the productive powers associated with it, he was advocating the goodness of the technological society.

Bhaskar's scientific realism already presupposed a duality between the transitive and intransitive dimensions to the extent that successes *must* be a consequence of the correct correspondence between human activity and the possibilities permitted by "the real". Success supposedly occurs as a result of the convergence between these two dimensions. It is for this reason that Bhaskar was able to argue that not only do the practical successes of experimental science demonstrate the validity of scientific realism, but also that facts have inherent value. However, if we posit that the intransitive dimension is emergent from the trajectories of the transitive labour process *as a totality*, then the teleological positings of labour provide the *poiesis* of agencies with a trajectory emergent from a whole history of past efforts and their satisfaction. The objects of any technological framework are organised according to their appropriateness for the *poiesis* of the labour process itself. Thus the productive possibilities of these objects are situated within the totality of the labour process according to the teleological positings associated with each and every object and the teleological positing of the whole labour process as a totality. The agential potentials and possibilities of any object should not be attributed to the object, on this account, but, instead, be regarded as properties of the way that they are situated within the organization of the whole labour process from beginning to end. These potentials and possibilities should not be divorced from the teleological positings of the labour process as a totality, and, without the organization of agencies within the whole labour process, the objects within that organization would have no potentials or possibilities whatsoever. As Lukács observed,

> Realization is not simply the real result that real men accomplish in struggle with reality itself in labour, but also what is ontologically new in social being in opposition to the simple changing of objects

in the processes of Nature. Real man, in labour, confronts the entire reality that is involved in his labour, and in this connection we should recall that we never conceive reality as simply one of modal categories, but rather as the ontological embodiment of their real totality.[20]

Thus labour processes are genuinely creative and transformative as they bring forth the unfolding of transformative powers along posited trajectories in accordance with the challenges put to labour. The act of "placing" the induced and abstracted *technai* of such processes in correspondence with an intransitive dimension of natural laws, that supposedly pre-exists the processes of labour, is an act of reification of those labour processes that alienates human beings from the reality which their labours bring-forth. This masks the social ontology of labour with an autonomous "objective world" of mechanised reality and places an obstacle to genuine inquiries into the ontology of labour itself.

Any genuine inquiry into labour processes should examine the teleology of those processes because without addressing the goals to which labour is destined, as well as the posited means by which those goals are to be satisfied, we cannot hope to understand those processes as a mode of organization of agencies. Since the sixteenth century, experimental physics has posited the form of truth through its mathematical projection of the six simple machines upon natural phenomena, whilst simultaneously positing the practical value of that truth for humanity. The organization of the ongoing activities of experimental research has transformed those natural phenomena in accordance with the posited anticipations of the form of truth and the human good life. Thus the reality disclosed by the labour processes of experimental physics should not be simply categorised as pre-scientific and objective Nature, but it should be seen as emergent from genuinely creative labour processes premised upon anticipations of truth and goodness. The agencies and transformative powers brought forth during experimentation should be taken to be the products of the totality, providing those products are situated within a historical trajectory, teleologically projected towards the discovery and liberation of power over our material conditions. Thus physics does not necessarily disclose a precedent reality but, rather, innovates and produces its own creative transformations of reality as sets, clusters, and ensembles of machine agencies and strata of transformative powers, in accordance with a moral equation that scientific truth is good. Physics is a consequence of the desire for certainty and technological power in a world that seems aloof, indifferent, and often hostile to human life. The societal gamble is emergent from the human struggle against and within chaotic competing powers beyond human control and comprehension. The gamble is that the world will become a better place because it will become more intelligible and human beings will become freer by becoming more powerful. Hence it is not merely a historical accident that natural philosophers and positivistic scientists have advocated a moral obligation and duty to pursue scientific truth. The assumption of an equation between technological and moral imperatives underpins the epistemological foundation for the techneic causal accounts that are emergent as abstracted and communicable understandings of those ensembles and strata of powers. Hence they are represented as the results of the successful reproduction of those labour processes and the cognition of their future possibilities. They are constructed in hindsight as a result of extending the closed system and removing all hindrances to its reproduction, whilst transforming all contact with the natural world into series of causal understandings juxtaposed with clusters of technological objects. Consequently, the functions and uses of the transformative powers produced by scientific activity do not remain fixed and defined at the walls of the laboratory and the pages of scientific literature (including journals, textbooks, and instruction manuals) as they are disseminated throughout the wider world. These functions are innovatively transformed and extended as they are implemented by the pluralistic agencies of the wider world. The social meaning and use of these objects are transformed as they are embedded and integrated within the cognitive, discursive, and technological practices of the everyday lived-world of modern life. They are transformed as they are utilised for political, commercial, and military purposes within the institutions and policies of states, corporations, and armed forces. These different dimensions of appropriation and usage are complex, pluralistic, intimately related, and frequently incommensurable. It is evidently the case that the experimental sciences are themselves merely instruments, technological objects, for the discovery and production of novel transformative powers for whatever purposes and challenges the wider world demands. Thus the possibility of the discovery of the final alethic truth of the transformative powers brought-forth by experimental physics is inextricably bound to the total complex labour processes of the final construction of the technological society.

However, the truth within the technological society is not only viewed in terms of its instrumental value for enhancing possibilities for producing an intelligible explanation of transformative power, but is perpetually deferred to the future because its realization is conceptually bound to its instrumental value in bringing-forth and explaining new transformative powers during the whole process of the construction of itself. Thus, the objective of this extension is not that of understanding the natural phenomena of the worlding of the world but is, rather, the creation of new labour processes, transformative powers, machine agencies, and the technological society. In this respect, the objective of experimental physics is its own self-creation, within a society that empowers that self-creation for its own purposes. Experimental physics is a means of disclosing the potentials and possibilities of itself as both an end-in-itself and as a means to future disclosures and refinements, within a society that values that metaphysical art as a source of trans-formative values and new powers. It is an art engaged in the *poiesis* of its own trajectories and destining, whilst it is embedded and inter-weaved within the social ontology of political, economic, and military agencies. Evidently, as I have endeavoured to argue above, it is ex-tremely problematical to describe physics as simply a process of the transformation of natural entities, via interventions and representa-tions, into knowledge. The mechanical realist conceptual grasp of phe-nomena as products of natural mechanisms utilises mechanical models as metaphors for the purpose of providing intelligible expressions and visualizations of phenomena as products. This consciously executed project promotes detachment and distancing of the subject-object rela-tion in the human reflections upon Nature. The construction of such metaphors and their intelligibility is an experimental process that is both challenged as the ongoing process of *Ge-stell* and is underwritten by the mechanical realist precepts. The construction of intelligible communicable accounts of novel phenomena is inextricably bound up with the processes of labour that produce those novel phenomena. During the construction of such accounts there is a continuous interac-tion between the labour processes involved in the innovation of novel communicative, representational, and material practices. It is a com-plex process of transforming the background of technological organ-ization and social organizations according to emergent teleological positings made to challenge and ontologically transform human-machine relations and agencies within a complex wider world of transformative powers. It is a "grand experiment"! Whether or not experimental physics could touch the asymptote of objective reality remains perpetually open to question because it remains incomplete and imperfect whilst it is endlessly challenged to test itself by innovating refinements and improvements of itself. The teleological positing of the mathematical projection perpetually defers the fulfilment in its own completion and perfection and the scientific realist has pre-empted the conclusions of the "grand experiment" by declaring that physics has achieved successes. If we examine the reality of the labour processes of experimental physics, from its historical origins to its contemporary trajectories, then, at most, we should limit our pronouncements of success to the more modest acceptance that the "grand experiment" is still ongoing. We have yet to determine whether the societal gamble of experimental science was a good move. This reveals the very real and ethically important realization that the societal gamble is a moral experiment. The scientific realist interpretation is not only an experimental metaphysical interpretation of the purpose of the possibility of human existence, but it also postulates norms regarding the nature of a good life. As such, it is a statement of allegiance and affirmation for the "grand experiment" that we call "the modern world".

No doubt, scientific realists will consider the argument in this book to be premised upon an idealist interpretation of science because they insist that the reality disclosed by scientific activities must be thought of as somehow being independent of human thought and action, whereas I have argued that the reality discovered by scientific activity is created by that activity. However, my argument is that it is "the scientific realist" who has superimposed an artificial construct over Nature and has attempted to construct a hegemonic moral posture as the defender of objective reality against pernicious idealism. Yet, perhaps, it is the so-called "scientific realist" who is the idealist – albeit a funny sort of idealist. The "scientific realist" insistence on the "independence" of their imagined conception of reality is something that not only is arbitrary, historically contingent, and rhetorical, but it is also potentially an epistemological obstacle to a deeper understanding of the nature of reality and the human condition. The abstract notion of "natural law" is an archaic mask that has been placed over the mystery that occurs during every act of making in the human world and every change in the natural world. This mystery is a mystery of "bringing forth" itself. Does a concept of "natural law" help us to understand this mystery? Does it say anything more than whatever happens was necessary? What is Nature? Is it what is necessary? Is it objective and law abiding? Is there a unified "it" at all? Perhaps the "scientific realist" has pre-empted and spoken for Nature with her/his human dreams, imagination, and hopes. Perhaps the objective reality of "Nature" is of a kind that is completely alien to our guesses about it and even beyond our comprehension altogether. Perhaps an aloof Nature merely demands our respect and awe. Any genuine realist must accept that possibility. However, in this book, I have been discussing the reality that is intimately bound up with scientific comprehension and technological innovation. It is that reality which is phenomeno-logically and conceptually dependent upon our participation. If we were to stop trying to explain reality in technological terms then that reality would cease being unfolded in that way. Human beings and our modes of being are only a part of the processes by which the ontology of the world is disclosed, whilst both reality and truth are bought-forth by the process of disclosing reality and truth in particular and situated ways. We do not control its unfolding because we are situated within an unfolding worlding of the world. This existential and alethic con-ception of ontology is distinct from the idealist identity of reality in "The Mind" or "The Absolute". It is also distinct from the all too human insistence that reality and truth are somehow "out there". Experimental physics is just one possible mode of disclosure amongst all the others. Music is another such mode of bringing-forth. It also occurs through the interactions between human interventions and machine performances, it is also an art directed towards its own perfec-tion, it is a *Ge-stell* which discloses itself as standing-reserve destining towards perpetual novelty, and its contours are also inscribed in terms of the artificial *technographe* of musical composition. The only differ-ence between music and physics, on my account, is that "realists" gen-erally do not attempt to explain how music (as distinct from mere sound) is possible in reference to "Nature". However, the creation and performance of music, as a bringing-forth, is also a truth. Not a truth in the sense of "correctness", although it can be precise, but a truth in the sense of *aletheia*. It discloses its own truth as a mode of disclosure. Music does not "correspond" to an external and objective reality. Nor does it exist only in the mind. The objective-subjective classification does not explain music. It does not explain physics either.

The philosophers of science have spent too long bound, with the shadow-puppeteers behind them, arguing about what they see in the shadows on the cave wall. They need to turn around, pay close atten-tion to the art of making shadow-puppets, and question how experi-mental physics is done. It has not only been my intention in this book to look closely at the shadow puppeteers to see how they relate the shadows to the art of making the puppets to produce shadows, but also to argue that the representations on the cave wall tell us nothing about the fire and the sun. If the experimenters are the shadow-puppeteers who learn their art of puppet making by watching the shadows on the cave wall change, as they make their shadow-puppets and pull their strings, then what is the fire? How does it relate to the sun? And, what is its origin? Physicists, positivists, scientific realists, and philosophers, need to attend to the question of the metaphysical foundation of experimental physics afresh, from within the context of a more general challenge that we all attend to *the phenomenon of making* as an existentially and metaphysically important phenomenon for any inquiry into the conditions and trajectories of the human character and our understanding of the world. This is crucial in our understanding of what the human character is and how we go about questioning it. Furthermore, we need to question the extent that the other experimental and natural sciences, such as chemistry and biology, are also bound up with mechanical realist metaphysics and the societal gamble. Hence, we need to reflect upon, debate, and question our deepest presuppositions and ideals about the ontological foundations of our possibilities when we derive our philosophical conceptions from the ideas, models, theories, and "facts" produced by the "natural sciences" regarding the human condition and our place within the world. Our understanding of the experimental "natural" sciences needs to be situated within a broader and deeper inquiry into how the human condition and reality is explored and changed through acts of making. Each experimental "natural" science is an innovative art that explores the questioning into its own possibilities by making them happen. Hence, the epistemology of the experimental sciences is derived from the same mechanical realist metaphysical assumptions regarding the ontological possibilities that permit acts of making. However, making, as one mode of being-in-the-world, is itself only as miraculous or surprising as any other mode of being-in-the-world. Once we have reached this level of truth then we are confronted with the reality that it is our existence that is surprising and miraculous. This reality is not explicable by scientific realism and, as a consequence, the successes of the experimental sciences are not explained by affirming realist metaphysics. The world remains surprising and miraculous however we attempt to explain it, because we need to explain why our explanation should be the case and then explain that further stratum of explanation. This stratified process of explaining could continue indefinitely, perpetually requiring further explanation, and, as I have already argued, the intelligibility of any explanation is independent from its truth-status. Explanatory realism only instrumentally functions within the ongoing process of stratification and is perpetually incomplete, and, as a consequence of this incompleteness, does not have any epistemological privilege in the face of the enduring mystery of Being. It can only conceal that mystery by pretending to have correctly pre-empted its logic of stratification. Whereas, as I have endeavoured to argue, realism is far from being the only position that does not make a "miracle" out of the successes of physics because it cannot fully explain the possibility of experimental physics at all. If it could then what need would we have for experimental physics as a route to truth?

At present we are unable to state certainties regarding the mode of being that we call "labour". We do not know the reality and possibilities of our own being-in-the-world from which labour as a mode of being-in-the-world springs. This "innocence" arises from our state of *thrownness* in the world. The societal gamble is that science and technology will improve this state of thrownness. We cannot even know the conditions under which we could even begin to control and comprehend "the worlding of the world" that is beyond us, no matter how much we try to harness and understand it in technological terms. It simply demands our respect. It is for this reason that I share Ellul's concerns about the way that the artificial world destroys and replaces the natural world and "the societal gamble" is a gamble on the superiority of an artificial world over the natural world. However, I disagree with Ellul's argument that the artificial world does not even allow the natural world to restore itself or enter into a symbolic relationship with it. In my view, the artificial and natural worlds are symbolically related within the technological society negatively in terms of the "power" of the artificial over the natural and the "freedom" of the technological society from the limits of the natural world. As I have argued above, modern experimental physics is only possible because its metaphysical presuppositions allow a series of ontological and epistemological presuppositions that have become tacitly embedded within teleological positing of the societal gamble to artificially construct a better world that contains new powers and freedoms. Mechanical realism allows that societal gamble to be naturalised, whilst normalising and structuring particular modes of human participation within it. As I consequence of this, my criticism of Bhaskar's conception of alethic truth is that it maintained an implicit connection with the more traditional correspondence notion and consequently superimposed an abstraction of the product of particular modes of labour (i.e. those involved in experimental physics) over all modes of labour as their possibility and condition, whilst it metaphysically naturalises the societal gamble and alienates human beings from the choices that gamble entails. Whilst I agree that there is nothing necessary or sufficient about the experiences of the lived-world when that world is one of hunger, poverty, disease, squalor, fear, and premature death, I wish to criticise the hegemonic social ontology implicit within Bhaskar's epistemological conception of alethic truth when he maintained the traditional reification of the labour processes of experimental work. He constrained and determined the possibility of human freedom in accordance with the abstracted outcomes of the labours of experimental scientists, whilst he presupposed that science is a good. However, the mechanical realist notions of rationality, truth, and progress are not easily sustained once we address the extent that technosciences are embedded in the ongoing challenges of *Ge-stell* and the societal gamble. The belief that the dissemination of the efficiency of technique and the power of technological objects will necessarily lead to human emancipation and enlightenment is naïve. The claim for the rationality, truth, and universality of the technoscientific enterprises of *Ge-stell* is simultaneously a claim for the legitimacy, power, and globalization of *Ge-stell*. Thus Bhaskar engaged in a hegemonic effort to pre-empt the shape of the future by declaring his allegiance to the authoritarian moral right of scientists and technologists to provide the means of production to shape that future in accordance with the entailments that their metaphysical abstraction has for all human knowledge, labour, and freedom. These entailments presuppose that rationality, truth, and freedom must be that which reasons and acts in accordance with the efficient and empowered participation in the technological imperative of the societal gamble. The unreflective presumption of mechanical realism alienates human beings from their own dialectical moral contemplation by attributing the duty of acting in accordance with the historical development of the technological imperative's drive towards efficiency and power as being the only possibility of achieving a free, rational, and good life for humanity. Once we take the societal gamble into account then we can also see that technosciences, such as modern experimental physics, entail moral presuppositions regarding the equations between power, freedom, truth, and the good.

Notes and References

1 Entering the Cave of the Shadow Puppeteers

1. B. Russell, *The Scientific Outlook* (Norton & Company, 1962), A.N. Whitehead, *Science and the Modern World* (Free Press, 1997), R. Carnap, *The Unity of Science* (M. Black (trans.), Thoemmes Press, 1995), K. Popper, *Conjectures and Refutations: The Growth of Scientific Knowledge* (Routledge, 2002) and *The Logic of Scientific Discovery* (London: Hutchinson, 1975). Also see G. Wolters and W.C. Salmon (eds), *Logic, Language, and the Structure of Scientific Theories: Proceedings of the Carnap-Reichenbach Centennial, University of Konstanz 21–24 May 1991* (University of Pittsburgh Press, 1994).
2. T.S. Kuhn, *The Structure of Scientific Revolutions* (Chicago University Press, 1962) and M. Foucault, *The Order of Things: An Archaeology of the Human Sciences* (New York: Vintage Books, 1994). Many of the themes raised by Foucault and Kuhn were anticipated in K. Manheim, *Ideology and Utopia* (Routledge and Kegan-Paul, 1936).
3. For examples, see D. Greenberg, *The Politics of Pure Science* (New York: New American Library, 1967), D. Bloor, *Knowledge and Social Imagery* (Routledge, 1976), M. Foucault, *Knowledge/Power: Selected Interviews and Other Writings, 1972–1977* (Gordon (trans., ed.), New York: Pantheon Books, 1980), B. Easlea, *Fathering the Unthinkable: Masculinity, Scientists, and the Nuclear Arms Race,* (London: Pluto Press, 1983), J. Harding (ed.), *Perspectives on Gender and Science* (London: Falmer Press, 1986), M. Ince, *The Politics of British Science* (Brighton: Wheatsheaf, 1986), D. Billig, *Arguing and Thinking* (Cambridge University Press, 1987), K. Danziger, *Constructing the Subject* (Cambridge University Press, 1990), P. Galison, *Image and Logic: A Material Culture of Microphysics* (Chicago University Press, 1997), and P. Galison and D. Stump (eds), *The Disunity of Science: Boundaries, Contexts and Power* (Stanford University Press, 1996).
4. For examples, see H-G. Gadamer, *Philosophical Hermeneutics* (University of California Press, 1966), H. Marcuse, *One-Dimensional Man: Studies in the Ideology of Advanced Industrial Society* (Boston: Beacon Press, 1966), M. Horkheimer, *Critique of Instrumental Reason: Lectures and Essays since the End of World War II* (O'Connell et al. (trans.), New York: Continuum, 1974), P.K. Feyerabend, *Against Method: Outline of an anarchistic theory of knowledge* (New Left Books, 1975), J. Baudrillard, *For a Critique of the Political Economy of the Sign* (Saint Louis: Telos Press, 1981), D. Haraway, *Primate Visions: Gender, Race, and Nature in the World of Modern Science* (New York: Routledge, 1989), T.P. Hughes, *Networks of Power* (Baltimore: John Hopkins Press, 1983), J. Habermas, *Knowledge and Human Interests* (Shapiro (trans.), Cambridge: Polity Press, 1987), D. Haraway, *Simians, Cyborgs, and Women: The Re-invention of Nature* (London: Free association Press, 1991), B. Latour, *We Have Never Been Modern* (Harvester Wheatsheaf, 1991), J-F. Lyotard, *The Postmodern Condition: A Report on Knowledge* (Bennington and Massumi (trans.), Manchester University Press, 1991), and T. Adorno, *The Stars Down*

to Earth, and Other Essays on the Irrational in Culture (Crook (ed.), London: Routledge, 1994).

5. For a detailed critical discussion of the neglect of experimentation within the traditional philosophy of science see A. Franklin, *The Neglect of Experiment* (Cambridge University Press, 1986).

6. For examples see B. Latour and S. Woolgar, *Laboratory Life* (Sage Publications, 1979), K. Knorr-Cetina, *The Manufacture Of Knowledge* (Pergamon Press, 1981), H.M. Collins, *Changing Order: Replication and Induction in Scientific Practice* (London: Sage Publications, 1985), P. Galison, *How Experiments End* (University of Chicago Press, 1987), B. Latour, *Science in Action* (Milton Keynes: Open University Press, 1987), L. Suchman, *Plans and Situated Actions* (Cambridge University Press, 1987), D. Gooding, T. Pinch, and S. Schaffer, (eds), *The Uses of Experiment: Studies in the Natural Sciences* (Cambridge University Press, 1989), D. Gooding, *Experiment and the Making of Meaning* (Dordrecht: Kluwer Academic Publishers, 1990), and A. Pickering (ed), *Science as Practice and Culture* (Chicago University Press, 1992).

7. M. Polanyi, *Personal Knowledge* (Routledge & Keegan Paul, 1958).

8. R.P. Crease, *The Play of Nature: Experimentation as Performance* (Bloomington: Indiana University Press, 1993).

9. As well as the traditional philosophers of science mentioned above, E.W. Strong (*Procedures and Metaphysics: A Study in the Philosophy of Mathematical-physical Science in the Sixteenth and Seventeenth Centuries*, Georg Ohms, 1966), R.S. Westfall (*Force in Newton's Physics: The Science of Dynamics in the Seventeenth Century*, New York: American Elsevier, 1971), A. Koyré (*Metaphysics and Measurement*, Maddison (trans.), Gordon and Breach, 1992), and L. Wolpert (*The Unnatural Nature of Science*, London: Faber and Faber, 1992), are all examples of philosophers of science for whom this derivative relationship between modern science and technology was self-evident.

10. A.N. Whitehead, *Science and the Modern World* (Free Press, 1997), p. 32.

11. E. Nagel, *The Structure of Science: Problems in the Logic of Scientific Explanations* (New York: Harcourt, Brace and World, 1961).

12. For examples see J.W. Cohen, "Technology and Philosophy" (in *Colorado Quarterly*, no. 4 (spring), 1955, pp. 409–420), H. Arendt, *The Human Condition* (University of Chicago Press, 1958), K. Marx and F. Engels, *Capital, vol. 1, A Critical Analysis of Capitalist Production* (Engels (ed.), Trans: More and Aveling, New York: International Publishers, 1967), P.T. Durbin, "Technology and Values: A Philosophical Perspective" (in *Technology and Culture*, 13, no. 4, October, 1972, pp. 556–576), L. Winner, *Autonomous Technology: Technics-out-of-Control as a Theme in Political Thought* (Cambridge: MIT Press, 1977), P.T. Durbin, "Towards a Social Philosophy of Technology" (in *Research in Philosophy and Technology*, 1978, pp. 67–97), D. Ihde, *Technics and Praxis: A Philosophy of Technology* (Boston: Reidel, 1979), L. Levidow and B. Young (eds), *Science, Technology, and the Labour Process* (London: CSE, 1981), J.K. Feibleman, *Technology and Reality* (London: Martinus Nijhoff Publishers, 1982), D. Ihde, *Existential Technics* (Albany: State University of New York Press, 1983), H. Jonas, *The Imperative of Responsibility: In Search of an Ethics for the Technological Age* (Chicago University Press, 1984), W. Bijker et al. (eds), *The Social Construction of Large*

Technological Systems (Cambridge Mass: MIT Press, 1987), D. Haraway, *Simians, Cyborgs, and Women: The Re-Invention of Nature* (London: Free Association Press, 1991), and B. Stiegler, *Technics and Time, 1: The Fault of Epimetheus* (Beardsworth and Collins (trans.), Stanford University Press, 1998).

13. For examples see H. Jonas, *Philosophical Essays: From Ancient Creed to Technological Man* (Englewood Cliffs, N.J: Prentice-Hall, 1974), P. Heelan, *Space-Perception and the Philosophy of Science* (Berkeley: University of California Press, 1983), D. Ihde, *Instrumental Realism: The Interface Between Philosophy of Science and Philosophy of Technology* (Bloomington: Indiana University Press, 1991), and C. Mitcham *Thinking Through Technology: The Path Between Engineering and Philosophy* (University of Chicago Press, 1994).

14. For examples see L. White, *Medieval Technology and Social Change* (Oxford University Press, 1962), P.K. Merton, *Science, Technology, and Society in Seventeenth Century England* (Cambridge University Press, 1970), P. Mathias (ed.), *Science and Society 1600–1900* (Cambridge University Press, 1972), J.A. Bennett, "The Mechanics' Philosophy and the Mechanical Philosophy" (in *History of Science*, 1986, pp. 1–28), S. Yearly, *Science, Technology, and Social Change* (London: Unwin Hyman, 1988), T.D. Kaufman, "Astronomy, Technology, Humanism, and Art at the Entry of Rudolf II into Vienna, 1577" (in *The Mastery of Nature: Aspects of Art, Science, and Humanism in the Renaissance*, Princeton University Press, 1993, pp. 136–50), and P.O. Long, "Power, Patronage, and the Authorship of Ars: From Mechanical Know-how to Mechanical Knowledge in the Last Scribal Age" (in *ISIS*, 88, 1997, pp. 1–41).

15. For example G.E. Marcus (ed.), *Technoscientific Imaginaries* (University of Chicago Press, 1986), S. Aronowitz, et al. (eds), *Technoscience and Cyber-culture* (Routledge, 1996), and A.J. Gordo-Lopez and I. Parker (eds), *Cyber-psychology* (London: Macmillan Press, 1999).

16. M. Heidegger, "The Question Concerning Technology" (in *The Question Concerning Technology and Other Essays*, Lovitt (trans.), Harper Torchbooks, 1977, pp. 3–35) and J. Ellul, *The Technological Society* (Wilkinson (trans.), New York: Knopf, 1964).

17. *QCT*, p. 14. Heidegger termed this mode of disclosure as *Herausforden*. A literal translation would be "to demand out hither". Lovitt noted (p. 14, fn. 13) that the verb *herausforden* could be translated as: to challenge, to call forth, to summon to action, to demand positively, or to provoke.

18. Lovitt noted (p. 15 fn. 14) that the verb *stellen* (to place or to set) has a variety of uses. It can be translated: to put in place, to order, to arrange, to furnish, or to supply. It can also be translated, in a military context, to challenge or to engage.

19. *ibid* p. 15.

20. Standing-reserve is a common translation for Heidegger's use of *Bestand*. Lovitt noted (*op cit* p. 17 fn. 16) that *Bestand* is ordinarily translated as "standing by" with its connotation of the verb *bestehen* ("to last" or "to undergo"). Lovitt also noted that *Bestand* contrasts with *Gegenstand* (object).

21. *ibid* p. 19.

22. *ibid* p. 25.

23. *Op cit* p. 32.
24. *The Technological Society*, "Note to the Reader", p. xxxv.
25. *ibid* pp. 11–19 and p. 79.
26. J. Wajcman, *Feminism Confronts Technology* (Cambridge: Polity Press, 1984) made this point about domestic "labour saving" devices. R. Wallis and S. Baran, *The Known World of Broadcast News* (Routledge, 1990) made this point about television news broadcasting. V. Walsh, "Contraception: The growth of a technology" (in L. Burke et al., (eds), *Alice Through the Microscope*, London: Virago, 1980, pp. 182–207) made this point about contraception.
27. See H. Simon, *The Sciences of the Analytical* (MIT Press, 1981) and M. Mueller, "Technology out of Control" (in *Critical Review*, 1.4, 1987, pp. 24–40) for detailed discussions of the idea of "technical rationality" as a context-dependent, evolving rationality.
28. Mueller, "Technology out of Control", p. 32.
29. *TS*, p. 125.
30. *ibid* pp. 6–12.
31. *ibid* p. 21.
32. *QCT*, p. 16.
33. *ibid* p. 14.
34. *ibid* p. 21.
35. *ibid* pp. 21–22.
36. *ibid* p. 23.
37. *TS* pp. 7–28.
38. *ibid* p. 86. See W. Chaloupka, *Knowing Nukes: The Politics and Culture of the Atom* (University of Minnesota Press, 1992), and Easlea, *Fathering the Unthinkable*, for discussions of the political and socio-technical imperatives implicit in the construction of the technosciences of atomic physics and nuclear power technologies.
39. *ibid* p. 88 and p. 103.
40. *ibid* p. 324 and p. 429.
41. *ibid* p. 5 and p. 21; for Ellul's discussions about the relationship between Nature, magic, and technique within primitive societies see pp. 24–7, 36–7, and 64–9.
42. Georg Lukács, *History and Class Consciousness* (Livingston (trans.), London: Merlin Press, 1967), p. 35, termed the individualism of the post Renaissance as "an individual consciousness à la Robinson Crusoe".
43. See Aristotle (N.E. Bk. 6) for his distinctions between the intellectual virtues of *episteme, techne, sophia, nous,* and *phronesis*. The criticism of the dominance of the intellectual virtue of *techne* in modern society was central to the critical analyses of modern society presented by Arendt, *The Human Condition*, J. Habermas, *Theory and Practice* (Viertel (trans.), Boston: Beacon, 1973), and J. Dunne, *Back to the Rough Ground: "Phronesis" and "Techne" in Modern Philosophy and in Aristotle* (Indiana: University of Notre Dame Press, 1993).
44. *QCT* pp. 6–13.
45. *Techne* τεχιη is commonly translated (see OED or the Liddle and Scott Greek-English Lexicon for example) from the Indo-European stem *tekhn-* ("woodwork" or "carpentry") as "art", "craft", "know-how", or "skill". The

Greek *tekton* and Sanskrit *taksan* are translated as "carpenter" or "builder". Sanskrit *taksati* is translated as "forms", "constructs", or "builds". The Hittite *takkss* – is translated as "to join" or "to build". The Latin *texere* is translated as "to weave".

46. I have used the following translations of the works of Plato and Aristotle: Plato, *The Complete Works* (Cooper (ed.), Hackett Publishing, 1997) and Aristotle, *The Complete Works* (2 volumes, Barnes (ed., trans.), Princeton University Press, 1984).

47. It is for this reason that Heidegger in his 1939 essay "On the essence and concept of *phusis* in Aristotle's *Physics* Book I" (in McNeill (ed.), *Martin Heidegger: Pathmarks*, Sheenan (trans.), Cambridge University Press, 1998, pp. 183–230), Kuhn, in *The Structure of Scientific Revolutions*, and Feyerabend, in *Against Method*, argued that Aristotelian and Galilean physics were describing two different things, using different methods of description, for different purposes, and therefore should not be empirically compared. In Kuhn and Feyerabend's terms, observations and measurements are theory-laden and there is not an impartial means by which they could be empirically tested and, consequently, the two physics were incommensurable. See also J. Bobik, "Matter and Individuation" (in *The Concept of Matter in Greek and Medieval Philosophy*, Mullin (ed.), University of Notre Dame Press, 1963, pp. 281–98).

48. "On the Essence and Concept of Phusis in Aristotle's Physics Book I" (in McNeill (ed.), *Martin Heidegger: Pathmarks*, Sheenan (trans.), Cambridge University Press, 1998, pp. 183–230), pp. 209–210.

49. According to Sambursky, *The Physical World of the Ancient Greeks* (Dugut (trans.), London: Routledge & Keegan-Paul, 1987) and R. Waterlow, *Nature, Change, and Agency, in Aristotle's Physics* (Oxford: Clarendon Press, 1982) there was a notable lack of experimentation in Ancient Greece for a period of over 800 years and that the works of Archimedes and the exploits of the legendary Daedalus were exceptional.

50. This destining finds its first and most explicit scientific expression in Norbert Weiner's studies (*Cybernetics or Control and Communication in the Animal and the Machine*, MIT Press, 1965) into the possibility of the science of cybernetics and the study if control systems. An example of one of the most extreme exponents of the mental transcendence of the human mind over technology is K.M. Sayre, *Cybernetics and the Philosophy of Mind* (International Library of Philosophy and Scientific Method, Routledge & Keegan Paul, 1976).

51. *Being and Time* (Macquarrie and Robinson (trans.), New York: Harper & Row, 1962). First published as *Sein und Zeit* in 1927.

52. "Letter on Humanism" (in *Basic Writings*, Farrel Krell (ed.), London: Routledge, 1999, pp. 217–265), p. 259, footnote.

53. "On the Essence and Concept of Phusis in Aristotle's Physics Book I", p. 220.

54. "The Age of the World Picture", p. 116, (in *The Question Concerning Technology and Other Essays*, pp. 115–54.)

55. *QCT*, pp. 21–3.

56. G. Lukács, *The Ontology of Social Being: 3: Labour* (London: Merlin Press, 1978) and *History and Class Consciousness*, pp. 83–109.

2 The Spirit of the Enterprise

1. For examples from contemporary experimental quantum physics see R.S. Cohen (ed.), *Experimental Metaphysics: Quantum Mechanical Studies for Abner Shimony*, vol.1 (Kluwer Academic Publishers, 1997).
2. I. Hacking, *Representing and Intervening: Introductory Topics in the Philosophy of Natural Science* (Cambridge University Press, 1983).
3. This is also apparent in the historical development of the sciences of psychology and artificial intelligence. See Jean-Pierre Dupuy, *The Mechanization of the Mind: On the Origins of Cognitive Science* (trans. DeBevoise, Princeton University Press, 2000) for an excellent discussion.
4. N. Maxwell, *The Comprehensibility of the Universe* (Oxford University Press, 1998).
5. *ibid* p. 5.
6. K. Popper, *Conjectures and Refutations: The Growth of Scientific Knowledge* (Routledge, 2002), pp. 192–3.
7. See R. Findlay-Hendry, "Realism and Progress: Why Scientists Should Be Realists" (in *Philosophy and Technology*, Fellows (ed.), Cambridge University Press, 1995, pp. 53–72), R. Harré, *Varieties of Realism* (Oxford: Basil Blackwood, 1986), R. Boyd, "The Current State of Scientific Realism" (in *Scientific Realism*, Leplin (ed.), Berkeley: University of California Press, 1984, pp. 41–82), and H. Putnam, *Realism and Reason* (*Philosophical* Papers, vol. 3, Cambridge University Press, 1985), for variants of this argument for scientific realism. Harré termed this kind of scientific realism as "policy realism".
8. R. Bhaskar, *A Realist Theory of Science* (Leeds Books, 1975).
9. *ibid* p. 24.
10. *ibid* p. 8.
11. *ibid* pp. 15–26.
12. This definition of an experiment was implicit to his argument in *A Realist Theory of Science* but he did not provide this explicit definition until later (*Scientific Realism and Human Emancipation*, London: Verso, 1986, p. 35).
13. *RTS* p. 227.
14. Popper, *The Logic of Scientific Discovery* (London: Hutchinson, 1975), pp. 106–111 and p. 423.
15. D. Gooding, *Experiment and the Making of Meaning* (Dordrecht: Kluwer Academic Publishers, 1990).
16. N. Cartwright, *How the Laws of Physics Lie* (Oxford: Clarendon Press, 1983).
17. *ibid* p. 62.
18. N. Cartwright, *The Dappled World: A Study of the Boundaries of Science* (Cambridge University Press, 1999).
19. *ibid* pp. 50–8.
20. B.C. van Fraassen, *The Scientific Image* (Oxford: Clarendon Press, 1980).
21. *ibid* pp. 19–40.
22. *ibid* p. 22.
23. For now, I shall put aside criticising the logical contradiction that we are able to make such a final statement about something we are supposedly unable to make any final statements about.
24. M. Merleau-Ponty made this point about the science of psychology in his criticisms of scientific accounts of perception in *Phenomenology of Perception*

(Smith (trans.), London: Routledge, 1999) and *The Visible and the Invisible* (Lefort (ed.), Lingis (trans.), Northwestern University Press, 1987). Heidegger (*Being and Time*) also made this point with regard to attempts to equate lived spatiality and temporality to any scientific conception of space and time. J.W. Dunne also made this point about scientific descriptions in general (*An Experiment With Time*, London: Faber & Faber, 1948, pp. 11–17).

25. J.K. Feibleman, *Technology and Reality* (London: Martinus Nijhoff Publishers, 1982), p. 6.

26. Gooding made this point in his discussion of the construction of novel communicable experiences in the early work on electromagnetism by Faraday et al. (*Measurement and the Making of Meaning*, pp. 29–30.)

27. As Popper observed (*The Logic of Scientific Discovery*, pp. 51–2), the positivist, by considering the problems of traditional philosophy as "metaphysical pseudo-problems" that are meaningless for empirical sciences, has neglected the fact that the central problem for traditional philosophy has always been a *critical analysis* of appeals to the authority of "experience".

28. *ibid* p. 206n and pp. 277–8.

29. *ibid* pp. 15–19.

30. See R. Harré and E.H. Madden, *Causal Powers* (Oxford: Blackwell, 1977), for a detailed discussion of this point.

31. "The Doctrine of Necessity Examined", p. 330 (in *Philosophical Writings of Peirce*, Buchler (ed.), Dover Publications: New York, pp. 324–38).

32. Kant was aware of the difficulties inherent in applying transcendental arguments to the experimental sciences. Kant too appreciated the difficulties in applying transcendental principles to the particularity of experiments in physics (and chemistry). The questions "How is physics possible?" and "How is the transition to physics possible?" were both central to his later work. These efforts were unpublished during his lifetime but have recently become available. Cf. Notes 22: 282 to 22: 452 in Kant, I., *Opus postumum*, Forster (ed.) Rosen & Forster (trans.), Cambridge University Press, 1995, pp. 100–199.

33. *RTS*, p. 21.

34. C. Norris, *Quantum Theory and the Flight from Realism*: *Philosophical Responses to Quantum Mechanics* (Routledge, 2000), especially pp. 48–9.

35. *RTS*, p. 114.

3   The Mathematical Projection of the Six Simple Machines

1. Galileo Galilei, *The Assayer* (in Drake (ed., trans.), *The Discoveries and Opinions of Galileo*, Garden City, N.Y.: Doubleday Anchor, 1975.)

2. M. Heidegger, "What is Metaphysics?", p. 83 (in *Basic Writings*, pp. 82–96.)

3. M. Heidegger, "Modern Science, Metaphysics, and Mathematics", p. 272 (in *Basic Writings*, pp. 267–306). First published in his 1962 published lecture *Die Frage nach dem Ding*.

4. For examples see G. Brett, "Byzantine Watermill" (in *Antiquity*, xiii, 1939, pp. 354–6), M. Rostovtzeveff, *Social and Economic History of the Hellenistic World* (Oxford, 1941), H. Chatly, H., 1942, "The Development of Mechanisms in Ancient China" (in *Engineering*, cliii, 1942, p. 145), M. Clagett &

A. Moody, *The Medieval Science of Weights* (Madison: University of Wisconsin Press, 1952), R. Dugas, *A History of Mechanics*, (Maddox translation, Routledge and Keegan Paul, 1955), R. J. Forbes, *Studies in Ancient Technology* (9 vols., Leiden: Brill, 1955), M. Clagett, *The Science of Mechanics in the Middle Ages* (Madison: University of Wisconsin Press, 1959), D.J. Price de Solla, "Ancient Greek Computer; with Biographical Sketch" (in *Scientific American*, June 1959), G. de Santilliana, *The Origins of Scientific Thought* (London: Weidenfield and Nicholson, 1961), White, *Medieval Technology and Social Change*, Wightman, *Science and the Renaissance*, A.P. Usher, *A History of Mechanical Inventions* (Harvard University Press, 1962), A.F. Burstall, *A History of Mechanical Engineering* (London: Faber & Faber, 1963), B. Gille, *The Renaissance Engineers* (London: Lund Humphries, 1964), H.F. Kearney (ed.), *Origins of the Scientific Revolution* (London: Longmans, 1964), H. Hodges, *Technology in the Ancient World* (Penguin Press, 1970), P.L. Rose and S. Drake, "The Psuedo-Aristotelian *Questions of Mechanics* in Renaissance Culture" (in *Studies in the Renaissance*, 18, 1971, pp. 65–104), M. Crosland (ed.), *The Emergence of Science in Western Europe* (Macmillan Press, 1975), P.L. Rose, *The Italian Renaissance of Mechanics* (Geneva: Droz, 1975), M. Clagett, *Archimedes in the Middle Ages*, 3 volumes (The American Philosophical Society, 1978), J.G. Landels, *Engineering in the Ancient World* (University of California Press, 1978), D.C. Lindberg, *Science in the Middle Ages* (Chicago University Press, 1978), E. Grant, *Much Ado About Nothing: Theories of Space and Vacuum from the Middle Ages to the Scientific Revolution* (Cambridge University Press, 1981), C.B. Schmitt, *Aristotle and the Renaissance* (Harvard University Press, 1983), D. Hill, *A History of Engineering in Classical and Medieval Times* (London: Croom Helm, 1984), W.A. Wallace, *Galileo and his Sources: The Heritage of the Collego Romano in Galileo's Science* (Princeton University Press, 1984), K.D. White, 1984, *Greek and Roman Technology* (Thames and Hudson, 1984), Bennett, "The Mechanics' Philosophy and the Mechanical Philosophy", W.R. Laird, "The Scope of Renaissance Mechanics" (*OSIRIS*, 1986, 2, pp. 43–68), D.L. Simms, "Archimedes and the Invention of Gunpowder" (*Technology and Culture*, 28, 1986, pp. 67–79), R. Sorabji, *Matter, Space, and Motion: Theories in Antiquity and Their Sequel* (Cornell University Press, 1988), W.R. Laird "Archimedes among the Humanists" (*ISIS*, 82, 1991, pp. 629–38), K. DeVries, *Medieval Military Technology* (Broadview Press, 1992), Long, "Power, Patronage, and the Authorship of Ars", and R.M. Torrance, *Encompassing Nature: Nature and Culture from Ancient Times to the Modern World* (Washington: Counterpoint, 1999). See also R. Temple, *The Crystal Sun* (Century Publications, 2000), for fascinating and controversial archaeological theories and evidence regarding ancient Egyptian, Hindu, Carthaginian, African, and Greek optics and astronomy.

5. L. White, *Medieval Technology and Social Change*, (Oxford University Press, 1962), p. 134.

6. See E.L. Eisenstein, *The Printing Press as an Agent for Change: Communications and Cultural Transformations in Early-Modern Europe* (Cambridge University Press, 1979) and Long, "Power, Patronage, and the Authorship of *Ars*" (in *ISIS*, 88, 1997, pp. 1–41), for detailed discussions.

7. J. A. Bennett, "The Mechanics' Philosophy and the Mechanical Philosophy" (in *History of Science*, 1986, pp. 1–28), p. 2.

8. The only scholarly certainty regarding the authorship of the works of Psuedo-Aristotle is that it probably was not Aristotle. They were possibly written by a student of Aristotle named Strato in the forth century BC. Also, as H. Cartelon argued, in "Does Aristotle Have A Mechanics?" (in Barnes et al., (eds), *Articles on Aristotle: Volume 1: Science*, London: Duckworth, 1975), it is impossible to construct a consistent mechanics from the few fragments on weight and motion scattered throughout the works of Aristotle. It is questionable whether the Aristotelian mechanics had anything more than a superficial and nominal connection with the philosophical writings of Aristotle. In fact, Aristotle argued that geometry could not provide a complete description of movement because it lacked a continuity axiom (*Physics* VI, 1, 231a24ff). The incompleteness of Euclidean geometry has been widely accepted by mathematicians and logicians since Frege published *Begriffsschrift* in 1879. Aristotle did permit mathematics a role in *technai* such as astronomy, optics, and music. See *Physics* II.2.194a7–11, *Post. Analytics*, I.7.75b14–20, and *Meteo.*, III.3–5. See also T.L. Heath, *Mathematics in Aristotle* (Oxford University Press, 1949).

9. See M. Clagett & A. Moody, *The Medieval Science of Weights* (Madison: University of Wisconsin Press, 1952), pp. 213–9, for details.

10. M. Clagett, *Archimedes in the Middle Ages* (The American Philosophical Society, 1978), p. 1225.

11. Plutarch, *Lives* (Perrin (trans.), Cambridge: Harvard University Press, 1961). Also see M. Authier, "Archimedes: The Scientists' Canon" (in *A History of Scientific Thought*, Serres (ed.), Blackwell Publications, 1995), for an interesting discussion of the narrative construction and use of Plutarch's legend of Archimedes.

12. Plutarch, *Lives*, pp. 4–5.

13. T.L. Heath (ed.), *The Works of Archimedes with the Method of Archimedes* (N.Y.: Dover, nd).

14. *The Method of Archimedes*, p. 13. Heiberg discovered this work in 1906. It was copied onto tenth century parchment (with only the final leaves written on sixteenth century paper).

15. See A.G. Keller, "Archimedean Hydrostatic Theorems and Salvage Operations in Sixteenth Century Venice" (in *Technology and Culture*, 1971, pp. 612–17), for a description of how salvage operations in Venice put Archimedes' principles into practice during the sixteenth century.

16. Clagett, *Archimedes in the Middle Ages*, pp. 490–1. See also D.L. Simms, "Archimedes' Weapons of War and Leonardo" (in *British Journal for the History of Science*, 21, 1988, pp. 195–210) for a description of Leonardo's design of the *Architronito* (a steam cannon) based upon the drawings of cannon in *De Re Militari* by Valturius, who stated that the cannon had been invented by Archimedes. There is a lack of any supporting evidence for Valturius' claim. Ioannis Sakas, a physicist, used Leonardo's sketches to build this device in the mid-1980s. Simms reported that it projected a missile (a 10 oz. tennis ball filled with hardened cement) to a distance of 150 to 200 ft. within seconds. See P. Gulluzi, *Leonardo da Vinci: Engineer and Architect* (Montreal Museum of Fine Arts, 1987, pp. 91–5) for a discussion of Leonardo's use of Aristotelian notions of motion (as presented in Pseudo-Aristotle's *Mechanica*) and the Archimedean principles of geometrical

mechanics. See M. Kemp, *Leonardo Da Vinci: The Marvellous Works of Man and Nature* (Harvard University Press, 1981) for a discussion of Leonardo's designs for machines and devices (including hydraulic devices, fortifications, weaponry, flying machines, submarines, the parachute, and the helicopter.)

17. W.C. Dampier, "From Aristotle to Galileo" (in Needham and Pagel (eds), *Background to Modern Science*, New York: Arno Press, 1975, pp. 32–3) argued that modern science could have begun with Leonardo Da Vinci if he had published his work.

18. Galileo, *On Motion and Mechanics* (trans. Drabkin and Drake, Madison: University of Wisconsin Press, 1960), p. 67.

19. A translated version of this paper, "The Little Balance", can be found in L. Fermi and G. Benadini, *Galileo and the Scientific Revolution* (New York: Basic Books, 1961).

20. This paper was unpublished until printed in 1638 as an appendix to the *Two New Sciences*.

21. This was unpublished until 1883.

22. Galileo *On Motion*, p. 38. In this work, he presented his "physical analogy" of the balance for naturally moving bodies and for the motion of a body on an inclined plane and a pendulum.

23. Published in French in 1634 and in Italian in 1649.

24. See T. Settle, "Galileo's Use of Experiment as a Tool of Investigation" (in McMullin (ed.), *Galileo: Man of Science*, New York: Basic Books, 1967, pp. 256–92). Also see R.H. Naylor, "Galileo's Experimental Discourse" (in Gooding et al., *The Uses of Experiment*, pp. 117–34) and G. Cantor, "The Rhetoric of Experiment" (in *The Uses of Experiment*, pp. 159–80) for discussions of Galileo's rhetorical use of his "experiments". Both Naylor and Cantor argued that Galileo anticipated the results of his experiments and question whether Galileo physically performed those "experiments" at all.

25. See R.J. Seegler, *Galileo Galilei: His Life and Works* (Oxford: Pergamon Press, 1966), p. 4.

26. Tartaglia's famous books are *Nova scienta* (1537) and *Questi et inventioni diverse* (1546). Translations of parts of *Questi* Bk.VII and all of Bk.VIII are in S. Drake and I.E Drabkin, *Mechanics in Sixteenth Century Italy* (University of Wisconsin Press, 1969) pp. 104–43.

27. See W.R. Laird, "The Scope of Renaissance Mechanics" (*OSIRIS*, 1986, 2, pp. 43–68), p. 52, and Clagett, *Archimedes in the Middle Ages*, pp. 508–607.

28. *Questi*. Bk. VII. For discussions of the Psuedo-Aristotelian basis of Tartaglia's mechanics see Laird, "The Scope of Renaissance Mechanics", pp. 52–3, and Wallace, *Galileo and his Sources Sources: The Heritage of the Collego Romano in Galileo's Science* (Princeton University Press, 1984), pp. 203–5.

29. *Questii*, fols. 78r–v, quoted and translated in Drake and Drabkin, *Mechanics in Sixteenth Century Italy*, pp. 105–7.

30. M.J. Osler, *Divine Will and the Mechanical Philosophy: Gassendi and Descartes on Contingency and Necessity in the Created World* (Cambridge University Press, 1994) p. 117. M. Mandelbaum, *Philosophy, Science, and Sense Perception: Historical and Critical Studies* (Baltimore: John Hopkins University Press, 1964).

31. *Questii*, fol. 82v, Drake and Drabkin, *Mechanics in Sixteenth Century Italy*, p. 111.

32. Laird, "The Scope of Renaissance Mechanics", pp. 53–5.
33. This book was published in 1613, well after Maurolico's death in 1575. The preface to this book and several of the problems can be found in Clagett, *Archimedes in the Middle Ages*, pp. 784–7.
34. Guidobaldo's famous books are *Le mechaniche* 1581 and *Liber mechanicorium*, published in Latin 1577 and Italian 1588, see Drake and Drabkin, *Mechanics in the Sixteenth Century Italy*, pp. 239–328. Also see Wallace, *Galileo and his Sources*, p. 206.
35. A.G. Keller, "Mathematics, Mechanics, and Experimental Machines in Northern Italy in the Sixteenth Century" (in M. Crosland (ed.), *The Emergence of Science in Western Europe*, Macmillan Press, 1975, pp. 15–34), pp. 21–32.
36. Laird, "The Scope of Renaissance Mechanics", pp. 56–57, and P.L. Rose, *The Italian Renaissance of Mechanics* (Geneva: Droz, 1975), pp. 248–51.
37. See C.B. Schmitt, "Science in the Italian Universities in the Sixteenth and Early Seventeenth Century", (in Crosland (ed.), *The Emergence of Science in Western Europe*, pp. 35–56) and *Aristotle and the Renaissance* (Harvard University Press, 1983). See also E. Grant, "Ways to Interpret the Terms 'Aristotelian' and 'Aristotelianism' in Medieval and Renaissance Natural Philosophy" (in *History of Science*, 25, 1987, pp. 333–58).
38. Tomeo was a professor of philosophy at Padua from 1497 to 1509. Cf. Laird, "The Scope of Renaissance Mechanics", pp. 48–49, and, P.L. Rose and S. Drake, "The Psuedo-Aristotelian *Questions of Mechanics* in Renaissance Culture" (in *Studies in the Renaissance*, 18: pp. 65–104), p. 79. He translated Pseudo-Aristotle into Latin and published a commentary in 1525. Piccolomini taught moral philosophy at Padua in 1539. Laird, p. 49, and Rose and Drake, p. 82. See R. Suter, "The Scientific Work of Alessandro Piccolomini" (in *ISIS*, 60, 1969, pp. 210–22) for an excellent discussion of Piccolomini's scientific work.
39. Laird, "The Scope of Renaissance Mechanics", pp. 60–2.
40. Moletti's arguments can be found in *In librium mechanicorum Aristotelis expositio tumultaria et ex tempore*. Milan, Biblioteca Ambrosiana, MS. S 100.
41. Published in 1620. Quotations and references are taken from the Jardine and Silverthorn edition (Cambridge University Press, 2000).
42. *New Organon*, pp. 6–7. Bacon did not give any examples of the arts, intellectual sciences, philosophy, or childish Greek sciences to which he referred. He only gave a single reference to Plato's reference to Atlantis in *Timaeus* (24D ff.) and the description of Scylla in Ovid (*Metamorphoses*, x. III. 732–3).
43. *ibid* pp. 20–1, my emphasis.
44. *ibid* pp. 69–70.
45. *ibid* p. 100.
46. *ibid* p. 28.
47. Laird, "The Scope of Renaissance Mechanics", pp. 65–67, and Wallace, *Galileo and his Sources*, pp. 208–216.
48. *On Motion and Mechanics*, p. 421.
49. *On Motion and Mechanics* on the pendulum pp. 152–3; on the astronomical sphere pp. 348–9; on steelyards and balances pp. 213–4.
50. On this point, I agree with E.W. Strong (*Procedures and Metaphysics: A Study in the Philosophy of Mathematical-physical Science in the Sixteenth and*

Seventeenth Centuries, Georg Ohms, 1966). However, Strong did not account for the use of machines or mechanical devices in the rise of mathematical natural science and, consequently, he could not explain this generalised principle of operation.

51. In *Two New Sciences*, p. 124, Galileo defined his notion of force (*forza*) in terms of the lever as "mechanical advantage". Galileo proposed the derivation of "most other mechanical devices" in terms of "the Law of the Lever", pp. 110–112, in reference to Archimedes' treatment of *Equilibrium* (cf. Heath, *The Works of Archimedes*, pp. 189–220).

52. P. Machammer, "Galileo's machines, his mathematics, and his experiments" (in *The Cambridge Companion to Galileo*, Machammer (ed.), Cambridge University Press, 1998) situated Galileo in "the Archimedean heritage" because the mathematical treatment of the balance was central to Galileo's physics.

53. *Two New Sciences*, pp. 252–6.

54. "Modern Science, Metaphysics, and Mathematics", p. 298.

55. "The Age of the World Picture", p. 127

56. For an English translation see Rene Descartes, *Rules for the Direction of the Mind* (Lafleur trans., Indianapolis: Library of the Liberal Arts, 1961).

57. Descartes, *Principles of Philosophy* (Miller and Miller (eds, trans.), Dordrecht: Reidel, 1983) p. 52, fn. 14. See the revised Adam and Tannery edition of *Oeuvres de Descartes* (Paris: Vrin/C.R.N.S., 1966–76, II, pp. 541–44).

58. *ibid* p. 285.

59. See Osler, *Divine Will and the Mechanical Philosophy*, ch. 5, and W.R. Shea, *The Magic of Numbers and Motion: René Descartes' Scientific Career* (Mass: Science History Publications, 1991) for further discussions of this point.

60. See É. Bréhier, "The Creation of the Eternal Truths in Descartes' System", (in *Descartes: A Collection of Critical Essays*, Doney (ed.), University of Notre Dame Press, 1968, pp. 192–208), E.M. Curley, "Descartes on the Creation of the Eternal Truths" (in *Philosophical Review*, 93, 1984, pp. 569–597), G. Hatfield, "Reason, Nature, and God in Descartes" (in *Science in Context*, 3, no. 1, 1989, pp. 175–201), and D. Garber, *Descartes' Metaphysical Physics* (University of Chicago Press, 1992), for further discussions and arguments.

61. Osler, *Divine Will and the Mechanical Philosophy*, ch. 1 and 5, Garber, *Descartes' Metaphysical Physics*, pp. 148–55, B. Rubidge, "Descartes' Meditations and Devotional Meditations" (in *Journal of the History of Ideas*, 51, 1990, pp. 22–49), pp. 27–29, and A. Funkenstein, *Theology and the Scientific Imagination from the Middle Ages to the Seventeenth Century* (Princeton University Press, 1986), pp. 179–192.

62. Osler, *Divine Will and the Mechanical Philosophy*, p. 127.

63. *Discourse on Method and The Meditations* (Sutcliffe (trans.), Penguin Books, 1968), p. 54.

64. *Principles*, pp. xxvi–xxvii and p. 85. See D. Garber, "Science and Certainty in Descartes", (in *Descartes: Critical and Interpretive Essays*, Hooker (ed.), John Hopkins University Press, 1978, pp. 114–151) and D.M. Clarke, *Descartes' Philosophy of Science* (Pennsylvania State University Press, 1982) for further discussion of this point.

65. *Discourse on Method*, p. 78.

66. *ibid* p. 91.

67. See S. Schaffer, "Glass works: Newton's prisms and the uses of experiment" (in Gooding *et. al.* (eds.), *The Uses of Experiment*, pp. 67–104) for a detailed discussion of this point.

68. For a brief discussion of Beekman's contribution to the mechanical natural philosophy, see R. Hooykaas, "Beekman, Isaac" (in *Dictionary of Scientific Bibliography*, Gillespie (ed.), 16 volumes, New York: Scribner, 1972), vol. I, p. 566. Beekman did not publish any major works but wrote philosophical letters to his contemporaries. For discussions of Gassendi's contribution, see T.M. Lennon, *The Battle of the Gods and Giants: The Legacies of Descartes and Gassendi, 1655–1715* (Princeton University Press, 1993), and Osler, *Divine Will and the Mechanical Philosophy*. Gassendi published *Syntagma philosophicum* in 1658. For discussions of Hobbes' contribution, see S.I. Mintz, *The Hunting of the Leviathan: Seventeenth Century Reactions to the Materialism and Moral Philosophy of Thomas Hobbes* (Cambridge University Press, 1969), and T.A. Spragen, *The Politics of Motion: The World of Thomas Hobbes* (University of Kentucky Press, 1973). Hobbes published the first part of *The Elements of Philosophy* in 1655. For discussions of Cavendish's contribution, see Mintz, *The Hunting of the Leviathan*, pp. 3–5, and R.H. Kargon, *Atomism in England from Hariot to Newton* (Oxford, 1966), ch. 7. For a discussion of Charleton's contribution, see L. Sharp, "Walter Charleston's Early Life, 1620–1659, and the Relationship to Natural Philosophy in mid-Seventeenth Century England" (in *Annals of Science*, 30, 1973, pp. 311–340). Charleton published *Physiologia Epicuro-Gassedo-Charltoniana, or a Fabrick of Science Natural Upon the Hypothesis of Atoms* in 1654. For a discussion of Digby's contribution, see M. Foster, "Sir Kenneth Digby (1603–1665) as Man of Religion and Thinker" (in *Downside Review*, 106, 1988, pp. 101–25). Digby published *Two Treatises. In the One of Which, The Nature of Bodies; in the Other, the Nature of Man's Soule; is Looked Into: In Way of Discovery, of the Immortality of Reasonable Soules* in 1644. For discussions of Mersenne's contribution, see P. Dear, *Mersenne and the Learning of the Schools* (Cornell University Press, 1988). Mersenne published *Quaestiones celebrimae in Genesim* in 1623, *L'impiété des déistes* in 1624, *La vérité des sciences* in 1625, and *Traité de harmonie universelle* in 1627. Descartes had met Mersenne, 1604 to 1609, at La Flèche during their education by Jesuits (see Dear, *Mersenne and the Learning of the Schools*, pp. 12–13, and Garber, *Descartes' Metaphysical Physics*, pp. 5–9, for further details). Descartes also frequently corresponded with Constantin Huygens (father of Christiaan) and Henry More (see Osler, *Divine Will and the Mechanical Philosophy*, pp. 118–119, for further details.)

69. For discussions of the influence of Descartes upon Boyle's experiments, see R.M., Sargent, *The Diffident Naturalist: Robert Boyle and the Philosophy of Experiment* (University of Chicago Press, 1995), and, S. Shapin and S. Schaffer, *Leviathan and the Air-Pump: Hobbes, Boyle, and the Experimental Life* (Princeton University Press, 1985). For discussions of the influence of Descartes on Huygens' scientific work, see Westfall, *Force in Newton's Physics*, chap. 4, A. Elzinga, *On a Research Program in Early Modern Physics* (New York: Humanities Press, 1972), and J.G. Yoder, *Unrolling Time: Christiaan Huygens and the Mathematization of Nature* (Cambridge University Press, 1988). See R.S. Westfall, "The Foundations of Newton's Philosophy of Nature" (in *The British Journal for the History of Science*, I,

1962, pp. 171–82) for a discussion of the influence of Descartes on Newton's early experiments.

70. See Newton's "Preface to the 1st Edition" of *Principia*, pp. xvii–xviii, (*The Mathematical Principles of Natural Philosophy and The System of the World*, 2 volumes, Cajori (ed., trans.), University of California Press, 1962).

71. *Principia*, pp. 398–9.

72. *Principia*, p. xvii.

73. *The Origin of Forms and Qualities According to the Corpuscular Philosophy* (1672) (vol. 3 of *The Works of the Honourable Robert Boyle*, Birch (ed.), 6 volumes, Hildesheim: Georg Olms, 1965), p. 13.

74. See Shapin and Shaffer, *Leviathan and the Air-Pump: Hobbes, Boyle, and the Experimental Life* (Princeton University Press, 1985), chap. 2, for a discussion of Boyle's arguments in favour of this.

75. R. Hooke, preface, *Micrographia* (London, 1665), quoted from Bennett, The Mechanics' Philosophy and the Mechanical Philosophy", p. 1. Also see R.S. Westfall, "Robert Hooke, mechanical technology, and scientific investigation" (in Burke (ed.) *The Uses of Science in the Age of Newton*, Berkeley, 1983, pp. 85–110).

76. For a discussion of the persistence of Aristotelianism throughout the Renaissance see E. Grant, "Aristotelianism and the Longevity of the Medieval World View" (in *History of Science*, 16, 1978, pp. 93–106).

77. See T.S. Kuhn, *The Copernican Revolution* (Harvard University Press, 1957) and R.S. Westman, "Three Responses to Copernican Theory: Johannes Praetorius, Tycho Brahe, and Michael Maestlin" (in *The Copernican Achievement*, Westman (ed.), University of California Press, 1975, pp. 285–345).

78. Long, in "Power, Patronage, and the Authorship of Ars", argued that there were close ties between the patronage of political elites and sixteenth century mechanists. White, *Medieval Technology and Social Change*, argued that the value of the new sciences for military, civic, and economic powers was central to their acceptance. Kaufman, in "Astronomy, Technology, Humanism, and Art at the Entry of Rudolf II into Vienna, 1577", argued that art, science, technology, and humanism were inter-related across disciplines and flourished in the circles of the Imperial court in Vienna of the sixteenth century. Bennet, "The Mechanics' Philosophy and the Mechanical Philosophy", argued that practical mathematics grew in England during the sixteenth and seventeenth centuries through the work of Dee, Hood, Davis, Recorde, Diggs, the Gresham College, Norman, Gilbert, Gellibrand, Briggs, et al., in developing techniques to attempt to solve problems of national import. See Merton, *Science, Technology, and Society in Seventeenth Century England*, for a detailed discussion of the close ties between Gresham College, the Royal Society, the English Navy, and the development of English mathematics in the early seventeenth century. Also see P. Rossi, *Philosophy, Technology, and the Arts in the Early Modern Era* (Attanasio (trans.), Nelson (ed.), New York: Harper & Row, 1970).

4   The "Making" of the Ground-Plan of Nature

1. T.S. Kuhn, *The Essential Tension* (Chicago University Press, 1977).

2. Osler, *Divine Will and the Mechanical Philosophy*, chap. 10.

3. I. Hacking, "Language, Truth, and Reason" (in *Rationality and Relativism*, Lukes and Hollis (eds), Oxford: Blackwell, 1982) and "'Style' for Historians and Philosophers" (in *Studies in History and Philosophy of Science*, 23, 1992, pp. 1–20).

4. J. Derrida, *Of Grammatology* (Spivak (trans.), Baltimore and London: John Hopkins University Press, 1967).

5. "The Age of the World Picture", pp. 128–34. Lovitt noted (p. 128 fn. 12) that *Weltbild* is conventionally translated as "conception of the world" or "philosophy of life" but the literal translation as "world picture" was more appropriate in the context of Heidegger's discussion. However, in this context, "conception of the world" also bears a close relation to the theme of Heidegger's discussion regarding "mathematical projection".

6. "The Age of the World Picture", p. 134.

7. For examples from contemporary Ultra Low Temperature quantum physics see S.N. Fisher et al., "The Force on a Wire Moving through Superfluid 3He-B" (in *Physica B*, 165 & 166, 1990) and "A Pumping Experiment in Dilute 3He-4He Solutions at Millikelvin Temperatures" (in *Journal of Low Temperature Physics*, 100, 1995).

8. *Measurement and the Making of Meaning*, p. 254.

9. Heidegger maintained the position expounded in "Modern Science, Mathematics and Metaphysics", as a starting point for his later essay "The Age of the World Picture". This essay should be read as Heidegger's development and "deeper" understanding of the experimental aspect of scientific research in the light of his "Question Concerning Technology" essay.

10. *Measurement and the Making of Meaning*, pp. 186–9.

11. H. Collins, *Changing Order: Replication and Induction in Scientific Practice* (London: Sage Publications, 1985), chap. 3.

12. "Age of the World Picture", p. 119.

13. Gooding, *Measurement and the Making of Meaning*, chap. 5, described how Faraday construed the movement of magnetic needles around electrical wires in terms of circular motion by inventing construals of the performance of the novel needle-wire-magnet experiments.

14. Gooding noted this (p. 63) in reference to Biot, Davy, and Faraday's experiments.

15. *ibid* pp. 78–80.

16. See Derrida, *Of Grammatology*, especially pp. 81–7. Latour & Woolgar, *Laboratory Life*, used Derrida's idea of *inscription* to characterise graph-plotting machines used in scientific work as *inscription devices*. Like Derrida, Latour & Woolgar treated science as a form of writing for which the aim was to produce text. They neglected to attend to the way that scientific inscriptions are fed-back into the processes of experimentation, as part of technique, in order to produce the inscription devices in the first place.

17. Euclid, "The Thirteen Books of *The Elements*" (in *Great Books of the Western World*, Heath (trans.), University of Chicago Press, 1952).

18. Cf. "Postscript" in the second edition (1970) of Kuhn's *The Structure of Scientific Revolutions*.

19. Cf. Heath, *Mathematics in Aristotle*.

20. Merleau-Ponty, *Phenomenology of Perception*, used the term *praktognosia* to denote the pre-conscious character of tacit and practical knowledge. It has

parallels with Polanyi's tacit knowledge except that, for Polanyi, *Personal Knowledge*, tacit knowledge was situated in the educated intellect whereas, for Merleau-Ponty, this knowledge was embodied in the situated and habitual motility of the existential body-subject.

21. I. Kant, *Critique of Pure Reason* (London: Macmillan, 1964).
22. *Measurement and the Making of Meaning*, p. 87.
23. *ibid* pp. 46–7.
24. *ibid* p. 117.
25. *ibid* pp. 122–3.
26. G. Deleuze and F. Guattari, *What is Philosophy?* (Burchell and Tomlinson (trans.), London-New York: Verso, 1994), chap. 5.
27. "Age of the World Picture".
28. *ibid* p. 120.
29. *ibid* pp. 123–5. Lovitt noted a translation difficulty in rendering *Betrieb* as "ongoing activity" (p. 124 fn. 10). He noted that *Betrieb* means the act of driving on, or industry, activity, as well as undertaking, pursuit, business, and can also mean management, or workshop, or factory. Heidegger qualified his use of *Betrieb* (pp. 138–9) as not intending any pejorative sense. He stated that he intended this word to convey and highlight that the industriousness of research only degenerates into "mere busyness" when, in the pursuing of its methodology, it closes itself from accomplishing novel projections of the ground-plan and, instead, takes that plan as given and simply accumulates results and calculations. (There is a parallel between Heidegger's use of *Betrieb* and Kuhn's use of the term "normal science".) Heidegger argued that "mere busyness" in scientific research was a consequence of the tendency of research to become completely dominated by industriousness rather than remain open to the ground plan that "gives the impression of a higher reality behind which the burrowing activity proper to research work is accomplished."
31. *ibid* p. 126.
32. *ibid* p. 127. Lovitt noted (fn. 11) that "set in place" is a translation of *gestellt*. The verb *stellen*, with its meanings of to set in place, to set upon, to challenge forth, and to supply, is fundamental in Heidegger's understanding of the modern age. Heidegger used it to characterise the manner in which science deals with the real and the related noun *Ge-stell* characterised and named the essence of modern technology.
33. "Science and Reflection" (in *The Question Concerning Technology and Other Essays*, pp. 155–182).
34. Reminiscent of Popper's use the metaphor of the net to describe the purpose of scientific theory to capture features of the world. Heidegger was aware of the capacity of nets to let things pass through as well as trap things.
35. *ibid* p. 163.
36. It also addresses the two-fold dimensionality of experimental science in Bhaskar's *RTS*. Changes in normative coherence corresponds to the "transitive dimension" whilst objectness invariance corresponds to the "intransitive dimension".
37. Lovitt noted (AWP, p. 167 fn. 21, and pp. 167–8 fn. 22) that "challenges" translates *herausforden* (lit. demands out hither). "Sets upon" translates

stellen (cf. *QCT* p. 15 fn. 4). Heidegger used the following verbs to characterise the conduct of modern science as theory: *nachstellen* ("to entrap"), *sicherstellen* ("to make secure"), *bestellen* ("to order" or "to command"), *festellen* ("to fix" or "to establish"), *vorstellen* ("to represent"), *umstellen* ("to encompass"), *erstellen* ("to set forth"), and *beistellen* ("to place in association with").

38. *ibid* pp. 170–1. Lovitt noted (fn. 25) that "state of affairs" translates the noun *Sachverhalt* but a more literal rending would be "relating or conjoining of matters".

39. *ibid* p. 169

40. *ibid* p. 173

5   The Anvil of Practice and the Art of Experimentation

1. *The New Organon*, p. 47, pp. 144–7, and pp. 180–1.

2. *Measurement and the Making of Meaning*, chap. 4

3. G. Morpurgo, *A Search for Quarks (a Modern Version of the Millikan Experiment): One Researcher's Personal Account* (Genoa: Mimeo, 1972), pp. 5–6.

4. See D.I. Bradley et al., "Potential Dark Matter Detector? The Detection of Low Energy Neutrons by Superfluid 3He" (in *Physics Review Letters*, 75, 1995) and C.B. Bäuerle et al., "Laboratory Simulation of Cosmic String Formation in the Early Universe using Superfluid 3He" (in *Nature*, 382, 1986) and "Superfluid 3He Simulation of Cosmic String Creation in the Early Universe"(in *Journal of Low Temperature Physics*, 110, 1998) for examples of published work on these models.

5. W.D. Hackmann, "Scientific Instruments: models of brass and aids to discovery" (in Gooding et al., *The Uses of Experiment*, pp. 31–66).

6. Aristotle termed arguments of this type as enthymemic arguments. See Aristotle's *The Art of Rhetoric*. These are similar to syllogistic arguments from which *episteme* could be reasoned, but instead of deducing the necessarily true these arguments construct the probably true. This construction involves the rhetorical skill of knowing how to persuasively manipulate one's audience and discourse in terms of their established prejudices and conventions. It was a *techne* and a counterpart of the syllogistic dialectic of philosophy.

7. Aristotle first coined and defined "metaphor" in terms of a displacement or movement from proper meaning in *The Poetics* 1457b6–9. See Paul Ricoeur, *The Rule of Metaphor* (Czerny et al. (trans.), Routledge & Keegan-Paul, 1987), chap. 2, for a fascinating and detailed discussion of this point.

8. See J. Richards (*The Philosophy of Rhetoric*, New York, 1965) for a good discussion of this point.

9. See M. Black, *Models and Metaphors* (Cornell University Press: New York, 1962), for his excellent discussion about the value of metaphors within novel scientific theories for the development of abstract and intelligible scientific language.

10. See E.H. Hutten, *The Language of Modern Physics* (London, 1958) for his famous argument for the necessity of metaphor for the development of novel physics.

11. N. Goodmann, *Ways of World-Making* (Hackett Publishing, 1978), p. 94.
12. R.W. Emerson, *English Traits* (Boston, 1860), p. 169.
13. N. Maxwell, *From Knowledge to Wisdom* (Oxford University Press, 1984).
14. Lukács termed this as "phantom objectivity" (*History and Class Consciousness*, p. 83). In this respect experimental physics is something of a paradox for a Marxist analysis because it can be both an exemplar of free-creative labour and the ultimate in alienation from the products of labour.
15. *TS*, p. 74.
16. *Technics and Time*, 1: *The Fault of Epimetheus* (Beardsworth and Collins (trans.), Stanford University Press, 1998), p. 113.
17. The etymology of artifice is from the latin *artifex*: one possessed of a specific skill. This derives from *ars* (skill) and *facere* (to make). [OED]
18. Of course, I am not suggesting that artifice is the only characteristic of human nature or character. Human nature also emerges through communicative narrative and social relations (i.e. love, trust, faith, laughter, happiness, authority, etc.). These can be transformed into techniques, but they can also be goods-in-themselves.

6   What Enables Us To Build Machines?

1. *The Mangle of Practice: Time, Agency, and Science* (University of Chicago Press, 1995).
2. *ibid* p. 69.
3. *ibid* p. 52.
4. *ibid* p. 51.
5. M. Krieger, *Doing Physics: How Physicists Take Hold of the World* (Bloomington: Indiana University Press, 1991).
6. *Mangle of Practice* p. 7.
7. *ibid* p. 16.
8. *ibid* p. 54.
9. *ibid* p. 69.
10. *ibid* pp. 188–91. See also A. Pickering, *Constructing Quarks: A Sociological History of Particle Physics* (Edinburgh University Press, 1987), sections 6.4 and 6.5.
11. *ibid* p. 102, fn. 25.
12. *ibid* pp. 111–12.
13. *ibid* p. 59 (my emphasis).
14. *ibid* p. 21.
15. *ibid* p. 79 (my emphasis).
16. *ibid* p. 74.
17. H.C. Oersted, "Experiments on the effect of a current of electricity on the magnetic needle" (in *Annals of Philosophy*, 16, 1820, pp. 259–276), p. 274.
18. The anthropologist Tim Ingold introduced this term to me at a conference in Oxford on Biocentrism vs. Anthropocentrism in 1995. Ingold coined this term to capture the way that certain humans, namely "deep ecologists", have circumscribed a region of the experienced world under the term "Nature" and criticise any position that does not respect the objectivity of that region for being "anthropocentric". It is an apt word to describe Bhaskar's position.

19. R. Bhaskar, *Scientific Realism and Human Emancipation* (London: Verso, 1986) and *Dialectic: The Pulse of Freedom* (London: Verso, 1996). This view is indebted to Engel's conception of human freedom. For example, see F. Engels' *Anti-Dühring* (London, 1969, pp. 136–137), first published in 1877, where he expressed the view that "freedom does not consist in any dreamt-of independence from natural laws; but in the knowledge of these laws whilst making them work towards definite ends".
20. *The Ontology of Social Being*, pp. 122–3.

Index